Papyrus and Wetlands Management in Africa

This book provides a comprehensive overview of the influence that the papyrus plant has had on African wetlands (with a particular focus on the Nile) in the past, and how this may change with the oncoming threats of climate change.

The papyrus plant is one of the fastest growing in the world and it forms the dominant vegetation in almost all sub-Saharan wetlands, and particularly the Nile Basin wetlands. Unlike other books on rivers and wetlands, this book focuses on this dominant and defining plant which plays such an important role in determining the structure and functioning of the Nile and other African wetland systems. It examines papyrus in the past, in the present and in the future, asking where and when did it evolve, what features have made it so widespread, productive and useful to humans, and how it has influenced the ecology of African wetlands. With over 250 million people relying on the water of the Nile to survive, the book examines the intimate relationship between, papyrus, people and the river. It concludes by discussing the current challenges to the sustainability of that relationship and whether this relationship can withstand the threat of changing climate and human demands as populations increase.

This book will be of great interest to students and scholars of wetland management, ecosystem services, water resource management, biogeography, sustainable development and African studies.

Michael B. Jones is a plant ecophysiologist and is an emeritus fellow in Trinity College Dublin and a member of the Royal Irish Academy. He is the author of *Miscanthus for Bioenergy Production* (Routledge, 2019).

Earthscan Studies in Water Resource Management

Flood Risk and Community Resilience
An Interdisciplinary Approach
Lindsey Jo McEwen

Water Politics and the On-Paper Hydropower Boom
Power, Corruption, and Sustainability in Emerging Economies
Özge Can Dogmus

Water Justice and Groundwater Subsidies in India
Equitable and Sustainable Access and Regulation
Gayathri D. Naik

Climate Change and Water Scarcity in the Middle East
A Transitional Approach
Marielle Snel, Nikolas Sorensen, and Reed Power

Urban Water Ecosystems in Africa and Asia
Challenges and Opportunities for Conservation and Restoration
Edited by Shamik Chakraborty, Amit Chatterjee, and Pankaj Kumar

Public Banks and Public Water in the Global South
Financing Options for Sustainable Development
Edited by Thomas Marois, David A. McDonald and Susan Spronk

Legal Protection of European Inland Waters
Solutions for Sustainable Management
Edited by Jarosław Dobkowski, Elżbieta Zębek and Jakub Jan Zięty

Papyrus and Wetlands Management in Africa
Past, Present and Future Challenges
Michael B. Jones

For more information about this series, please visit: www.routledge.com/Earthscan-Studies-in-Water-Resource-Management/book-series/ECWRM

Papyrus and Wetlands Management in Africa

Past, Present and Future Challenges

Michael B. Jones

Routledge
Taylor & Francis Group
LONDON AND NEW YORK

earthscan
from Routledge

First published 2026
by Routledge
4 Park Square, Milton Park, Abingdon, Oxon OX14 4RN

and by Routledge
605 Third Avenue, New York, NY 10158

Routledge is an imprint of the Taylor & Francis Group, an informa business

British Library Cataloguing-in-Publication Data
A catalogue record for this book is available from the British Library

ISBN: 978-0-367-67498-4 (hbk)
ISBN: 978-0-367-67500-4 (pbk)
ISBN: 978-1-003-13154-0 (ebk)

DOI: 10.4324/9781003131540

Typeset in Optima
by Taylor & Francis Books

Contents

Figures

Preface

The main subject of this book is the papyrus plant, which has the Latin name Cyperus papyrus, and is probably best known to us as the plant used by ancient Egyptian cultures to make the first examples of paper, which they also called papyrus. Not surprisingly the word 'papyrus' is generally accepted as being of Egyptian origin. Papyrus is the Greek name of the plant and may come from the Egyptian word *papuro* meaning the 'the royal' or 'that of the Pharaoh' because the central government had control of the papyrus processing as they owned the land and oversaw the cultivation of the plant. The use of the word papyrus for the plant as well as the paper made from it makes the use of the word confusing, but when I use the name papyrus I usually mean the plant and not the paper.

There is plenty of evidence that the papyrus plant was abundant and widely distributed in Egypt at the time of the Pharaohs, but by the time that Napoleon's army invaded Egypt in 1798 it was almost extinct. The first time I saw papyrus in its natural environment was when I flew over the margins of Lake Victoria to land at Entebbe airport in Uganda. This was in 1971 and I was about to take up my first job after completing my PhD at Lancaster University in the north of England. I had just been appointed as a lecturer in botany at Makerere University in Kampala, the capital of Uganda. I was about to join a small group of colleagues lead by ecologists John Gaudet and Keith Thompson in the Botany Department at Makerere University, who had recently decided to take an interest in the papyrus wetlands, and in particular in the ecology of papyrus which grew in abundance around the perimeter of Lake Victoria and in many of the flooded valleys which eventually flowed into the lake. I did not see myself as a plant ecologist, but rather a plant physiologist who had spent my time for my PhD studying plants in the laboratory and trying to fathom out how they functioned. In fact, my PhD had been on a relatively obscure topic at the time; the study of what are referred to as 'circadian rhythms' of a form of photosynthesis in succulent plants called crassulacean acid metabolism (CAM). Circadian rhythms are innate daily cycles of processes that occur in all living organisms that are regulated by an internal 'clock'. It had been shown that succulent plants take up carbon dioxide at night, but that they do this, showing peaks and troughs on a twenty four hour cycle, even when kept in continuous darkness. Along with my supervisor Terry Mansfield, we concluded that this must be controlled by the internal clock, but what are the controls? This was an interesting topic to investigate more thoroughly, but detailed knowledge of this internal clock was to await the later development and application of the expanding field of molecular biology. The research would also be laboratory based and require facilities beyond those available to me in the university in Uganda. Therefore, given the opportunity for me to explore my exciting new environment in Africa, I decided it was now the time to take some of my laboratory-based studies, particularly related to the physiology of photosynthesis, into the field, and combine ecology and physiology in what was a relatively new but rapidly developing interdisciplinary field of ecophysiology. The aim of the studies was to describe the physiological functioning of papyrus wetlands and to understand the role that papyrus plays in the maintenance of these important ecosystems. This was crucially important as, even then, the papyrus-dominated wetlands were seen as increasingly impacted by human activities, and the papyrus plants, as an important component of the ecosystem, were particularly threatened and could possibly become extinct in large parts of Africa – a fate that had already occurred in most of Egypt many centuries ago.

Initially, I collaborated with Robert Milburn, the then Professor of Botany at Makerere University, to investigate the photosynthetic characteristics of papyrus to show, for the first time, that it has a highly efficient form of photosynthesis, which helped to explain the plants very high growth rate. These studies required instruments that were bulky and difficult to move, so the plant was taken into the laboratory and it took some time before totally portable instruments were able to measure carbon dioxide uptake by photosynthesis in the field. Their arrival in the late 1970s, revolutionised field studies of photosynthesis and by this time, now based in Trinity College Dublin, I was able to undertake field studies of papyrus wetlands in Rwanda, Uganda and Kenya. Over the years I have benefitted greatly from enlightened financial support from the Irish Department of Foreign Affairs aid programmes and from various EU Research Framework Programmes as well as the hospitality of the Institute of Environmental and Natural Resources, Makerere University, Kampala, Uganda. I would like to express my thanks to my fellow researchers who have accompanied and supported me on my journeys with papyrus wetlands, which are increasingly difficult to navigate as you get older! They include Frank Kansiime, Frank Muthuri, Frances Kariuki, Alfonse Opio, Steve Humphries and Matt Saunders. I also owe an especial debt to my wife, Sue, who has continually encouraged me throughout my career and particularly in the writing of this book.

1 An Introduction to Papyrus and its Environment

Tropical Wetlands

The places where papyrus grows are wetlands, which are permanently or seasonally inundated freshwater habitats ranging from lakes and rivers to marshes and swamps. It has long been proposed that tropical wetlands were closely linked with human origins of populations and their initial migrations. Chan et al. (2019) suggest that the earliest human line emerged within the paleo-wetlands of southern Africa approximately 200,000 years ago. There was a 70,000 year long existence in this region before an out of homeland northern–southwest dispersal 130,000–110,000 years ago. This was probably a result of climate change that increased humidity and opened a green corridor first northeast, then to the southwest.

Wetland ecosystems occupy the transitional zone between permanently wet deep-water bodies and generally dry areas. The most recent estimate of global inland and coastal wetland area is in excess of 12.1 million km^2, an area larger than Canada. Of this, about half of the area is permanently inundated, while an estimated further 5.2 million km^2 are intermittently or occasionally inundated. The largest areas of wetlands are in Asia (32% of the global area), North America (27%) and Latin America and the Caribbean (16%). Other areas of wetland are in Europe (13%), Africa (10%) and Oceania (3%). Li et al. (2022) used Landsat satellite images produced the first systematic and comprehensive map of existing wetlands in Africa that quantified the area and distribution of wetlands at a spatial resolution of 30 m. Their results showed that the existing wetland extent in 2020 was approximately 1.34 million km^2 comprising 3.98% of Africa's total land area. Of these wetlands, inland wetlands account for 81%, and coastal wetlands account for 19%. The wetlands in Africa were systematically distributed around the equator and decease with increasing latitude.

Wetlands have been called the 'kidneys of the earth' (Mitch and Gosselink, 2007) because of the role they play in enhancing the quality of water that flows through them. They also help to mitigate floods by slowing runoff from the land, protect coastlines and store and sequester carbon. The major

DOI: 10.4324/9781003131540-1

types of tropical freshwater wetlands are floating swamps (referred to as 'sudd' in East Africa), rooted permanent and seasonal herbaceous swamps, swamp forests and floodplains. In Africa the term 'swamp' generally applies to wet areas covered with tall, herbaceous vegetation which can be floating or bottom-rooted (Howard-Williams, 1977; Howard-Williams and Gaudet, 1985; Thompson, 1985). Floating swamps are more common in the tropics than elsewhere, and in Africa examples of the dominant plant species are the grass *Vosia cuspidata* and the sedge *Cyperus papyrus*. Freshwater wetlands tend to be highly productive ecosystems as they have ready access to water and nutrients from either the sediments or water column. Nutrient supply to tropical wetlands may be seasonal with higher inputs resulting from rainy season inflows. The water-saturated soils and sediments of wetlands are depleted in oxygen (they are referred to as anaerobic), and this has a critical impact on the biological, physical and chemical characteristics of the ecosystem that they form. Plants that grow in wetlands are referred to as hydrophytes, because they are specially adapted to grow with their roots in anaerobic environments (Osborne, 2000).

In recent years, tropical wetlands have received increased attention because we have now recognised their importance as large carbon stores, in the form of peat, that has frequently accumulated below the living vegetation over many years. The main areas of the most active peat accumulation are located in central Africa, south and central America, and southeastern Asia (Page et al., 2011; Gumbricht et al., 2017; Ribeiro et al., 2021; Dargie et al., 2017). The current interest is because it is now clear that peatlands can act as a significant net sink for atmospheric CO_2 fixed in photosynthesis and therefore have an important influence on the global carbon budget and the rate that CO_2 accumulates in the atmosphere. Accurately determining the amount of carbon stored in these peatlands depends on reliable measures of peatland area, depth of peat deposits, and carbon density. To date, there have been relatively few detailed estimates of peat depth, particularly in the more remote areas of Africa and South America. Based on new more reliable peatland area estimates made using satellite-based measurements and limited direct sampling of peat depth, Ribeiro et al. (2021) have estimated that tropical wetlands store 152–288 Gt of carbon, or about half the total global peatland stocks. However, as well as storing carbon, and therefore drawing down a greenhouse gas in the form of CO_2 from the atmosphere, wetlands are a source of carbon in the form of methane, another but more powerful greenhouse gas, released to the atmosphere under the anaerobic conditions in the peat. The net balance of these greenhouse gases between sinks and sources will determine whether wetlands have a significant role in regulating global warming caused by greenhouse gas accumulation in the atmosphere. This is an issue which will be taken up and expanded on in Chapter 4.

Papyrus Distribution

Although papyrus is currently found throughout the African continent, it has a disjunct distribution associated with the wetland areas in the major river catchments. The main centres of papyrus distribution include the swamps in the Sudd of South Sudan, the Okavango inland delta of Botswana, the Upemba region in the Congo including Lake Turba, Mai Ndombe and the complex of wetlands bordering the lower Ubangi and the Congo rivers, and the Bagwaulu and Kafue Flats along the Zambeze. It also grows along the shores of most freshwater lakes in East Africa, the largest of which is Lake Victoria with extensive fringing papyrus swamps. Large areas of papyrus have also been recorded on Lake Chad, which is an outlier in west/central Africa, but this lake and the area occupied by papyrus has shrunk dramatically in recent years. More details of papyrus distribution can be found in *A Directory of African Wetlands* by Hughes and Hughes (1992).

Migahid (1947 and 1948) published some of the more detailed vegetation surveys of the present day Nile Region and identified the main species of the permanent swamps as *Cyperus papyrus, Vossia cuspidata, Phragmites communis* and *Typha australis*. He was also the first to identify the main controllers of their distribution as water depth and current velocity. He distinguished between species such as *C. papyrus* and *V. cuspidata*, with large rhizomes that can float on the surface of the water as it rises and falls, and species like *Phragmites* and *Typha*, which are anchored in the substrate by their roots and are unable to survive deep flooding as water levels rise. Both depth and duration of flooding are important controllers of distribution. Unlike *Phragmites* roots, papyrus roots do not penetrate far into the substrate. Of course, this means that they may be vulnerable to being swept away by water currents and this is often the fate of floating islands of papyrus that can become detached in fast flowing water or during storms. Sutcliffe and Parks (1999) in a review of the Nile hydrology suggested that there is a distinct boundary between the shallow and the deep-water floating species controlled by a maximum depth of flooding at about 1.3 m. The most productive and visually vigorous papyrus forms a free-floating zone but it can also survive with its roots penetrating the soil both where the soil is saturated year-round but also in seasonally flooded zones, which can be dry during part of the year. Estimates of the geographical distribution and area of papyrus swamps in Africa have been difficult to determine precisely. There have been a few inventories of major wetlands published, but collating this information to identify the areas occupied by papyrus and not by other wetland vegetation such as *Typha* has proved difficult (Adam and Mutanga, 2009). Perhaps the most reliable estimate comes from Hughes and Hughes (1992), who published a comprehensive review of the state of African wetlands in the 1980s and at that time estimated the total area occupied by papyrus to be 40,000 km^2. However, since then we know that many wetlands have been degraded so that the total area has declined significantly. As we have seen, major wetlands where papyrus is commonly found occur mainly in East Africa and in much of this area human-induced change has led to widespread degradation of the papyrus dominated wetlands, particularly around Lake Victoria. The overriding indirect driver behind this is population growth. The population of sub-Sahara Africa is expected to grow from about 1.3 billion at present to around 2.5 billion in 2050. Increasing demand for food from the expanding population has put pressure on farmers to drain wetlands for large scale farming (van Dam et al., 2011). Much of this has been the result of corporate or government activity, most frequently for the cultivation of rice (see Chapter 5).

In many parts of the Nile catchment, swamps in which papyrus is dominant are the most widespread of the types of wetland vegetation. Probably the first modern-day description of the ecology of papyrus was published by Carter (1955) based on a year-long survey of papyrus swamps, mainly in Uganda at a point where the Nile leaves Lake Victoria. In Uganda and around Lake Victoria these littoral shallow water swamps dominated by papyrus are very widespread around the perimeter of lakes, the largest of which is Lake Kioga, in central Uganda, with an area of approximately 2,000 km^2 covered by papyrus. Carter (1955) also surveyed a papyrus swamp typical of those that commonly block valleys in western Uganda.

The lake-shore swamps have a succession of vegetation zones between the open water and the shore. From the open water the outermost zone consists of floating water-lilies (*Nymphaea caerulea* and *N. lotus*). This is followed by the sedge and fern zone, and then the floating papyrus zone. The papyrus zone is normally far broader than the other zones and papyrus is dominant with almost complete exclusion of other species. Finally, on the shore-side there is usually a narrow zone composed of mainly sedges and grasses, including *Miscanthidium violaceum* and *Loudetia phragmatoides*, which pass directly to the grassland of the shore (Lind and Visser, 1962). Lake levels fluctuate seasonally and over decades due to changes in rainfall locally and regional changes in inflow. For example, the surface area of Lake Naivasha in Kenya (approx. 150 km^2) fluctuates by 30% between periods of high and low rainfall and in the longer term, the lake level has varied by at least 9 m since 1910 (Boar, 2006). Gaudet (1977a) described the effects of lake level fluctuations on Lake Naivasha, Kenya. This elucidated some of the factors affecting lake-edge succession, and concluded that of all the species present, papyrus is the most adaptable, starting from seed germination and seedling growth on bare mud in competition with flooding tolerant annuals and later growing up in water as a perennial emergent macrophyte. It was also demonstrated that the production of bands of papyrus swamp along the lake edge was correlated with the irregular cycles of drying and flooding which have occurred on Lake Naivasha.

The papyrus swamps of the valleys are very similar in their structure to the papyrus zone of the lake swamps, except that in most places in the valley the mat of papyrus roots and rhizomes lies on the mud and is not floating.

However, except in very dry weather, there is water between the culms and this rises in wet weather. This water generally flows as a slow current down the swamp towards the valley outlet although there can be a central channel of open water where there is faster flowing water. Carter (1955) distinguished between two types of water in these swamps; lake water, largely derived from rain falling directly on the surface of the lake and land water, flowing off the land into the swamp after rain. The water from the two sources differs in many of their chemical characteristics and in particular the high levels of oxidised iron which give a red colour to the land derived water making it easily distinguishable from pure rainwater. In the valley swamps, the shore part of the of the papyrus zone in the lakes and below the mat of papyrus roots the water is highly de-oxygenated and has a high CO_2 content. Under these conditions the sediments in these regions produce large amounts of methane gas.

Papyrus – the Plant

Taxonomically, papyrus is a sedge and belongs to the family *Cyperaceae*. The *Cyperaceae* are grass-like, herbaceous plants found especially in wet regions throughout the world. The family comprises about 70 genera and 4,000 species. *Cyperus papyrus* (papyrus) is one of about 600 species in the genus *Cyperus*.

Papyrus is a plant species that has evolved to exploit an environment which is permanently water-flooded and in a tropical (warm) climate. These conditions are optimal for plant growth and as a result papyrus is among the fastest growing plant species in the world, producing in aerial growth more than 20 g DM m^{-2} day^{-1} (Thompson et al., 1979; Jones, 1987; Muthuri et al., 1989; Jones and Muthuri, 1997; Saunders et al., 2014). But why is it so productive, outstripping most other tropical plants? There are a number of anatomical and physiological characteristics of the plant, which we refer to as traits, that contribute jointly to its high productivity. Possibly the most important of these is the use of an efficient form of photosynthesis, utilising what is known as the C_4 pathway, to assimilate CO_2, but the structural characteristic of the canopy that intercepts most of the incident sunlight is also crucial (Jones and Milburn, 1978; Jones and Muthuri, 1985). In addition the plant is able to obtain a supply of organic nitrogen from nitrogen gas in the atmosphere in the process of nitrogen fixation by microorganisms associated with its roots and rhizomes. The combination of efficient photosynthesis and nitrogen fixation, along with a continuous supply of water in wetlands and a tropical climate makes for record growth rates. As a result papyrus is able to out-compete most other plant species in wetland environments and form a virtual monoculture over vast area of tropical wetland.

Papyrus is almost unique among plant species in being depicted in some detail by the early occupants of wetlands as long ago as 3,000 BCE. These records were made by the ancient Egyptians who had settled in the Nile

Valley and had established a civilisation that was to last three millennia. Papyrus was inextricably associated with the ancient Egyptians in being used to make the medium on which many of the early records were made and it is from the records on surviving papyrus scrolls that we have gained much of our knowledge about the land of the pharaohs. When drawing, whether on papyrus sheets or on the walls of tombs and temples, ancient Egypt's artists illustrated their subjects as clearly as possible using strict rules that showed images in a combination of profile, semi-profile and frontal view. These illustrations are so good that they have even allowed zoologists today to identify the exact species of birds and animals that lived on the banks of the Nile. The earliest surviving extensive writings about papyrus were made by the Greek philosopher Theophrastus (371–287 BCE) who has been given the title of 'Father of Botany'. The most important of his writings are two large botanical treatises titled *Enquiry into Plants* and *On the Causes of Plants*.

For the earliest settlers of the Nile Valley, looking for access to water, the papyrus swamps were probably seen as a threat as the mass of vegetation impeded navigation and the swamps were a source of danger in the form of insects that carried diseases and probably large animals, including the particularly aggressive hippopotamus. Furthermore, the annual cycles of flooding of the swamps and the surrounding floodplain meant that the land, particularly in the Nile Delta, could not be settled permanently. These floodplains had deposited on them annually silt carried by the river which provided a fertile environment for growing crops but also annual grasslands for grazing animals. Over time the areas that flooded were controlled by digging drainage channels and permanent settlements were established on raised banks that were beyond the reach of the annual floods. But in the meantime it was possible to occupy the floating papyrus swamps which rise and fall with the water levels. This floating world for the earliest settlers was dependent on fish and fowl for their basic foods and they lived in huts constructed from papyrus culms (Gaudet, 2014). These reed huts were in common use before 7,000 BCE, and they were not exclusive to the swamp dwellers. It is believed that most people in those days used papyrus to construct housing. They were built on land or papyrus rafts and even large boats, which are depicted in mural scenes in dynastic tombs and temples. Eventually the aquatic lifestyle largely gave way to settlement on solid ground of the pharaonic Egypt when cultivation for crops became far more intensive to support the increasing population. Houses were made from mud brick from before 3,200 BCE in the earliest dynastic period as organised agricultural settlements were established but the use of papyrus rope was still retained for tying together the wooden beam supports. The early settlers in the predynastic period also found a range of uses for the papyrus which were essential for their survival and everyday needs. Light boats or skiffs for navigating the waterways particularly during the Nile floods were the most important initially, but alongside this, ropes were essential for the construction of boats and buildings, as well as moving heavy objects.

Papyrus played a central role in the development of ancient Egyptian culture. It was a multifunctional material which impacted the daily life of ancient Egyptians. It frequently appeared in paintings, sculptures and murals as a symbol of fertility, protection and abundance. It influenced art, literature and daily life. The use of papyrus in artistic representations was a visual reminder of its importance in many aspects of everyday life. It was a symbol of abundance and prosperity because of its widespread availability and practical uses. It also symbolised protection as the swamps provided shelter for birds and small animal life. At the same time the dark and mysterious nature of papyrus swamps was used as a dominant element in Egyptian mythology.

Papyrus swamps were the settings or some of the most important of the ancient religious legends of the Egyptians. In ancient Egyptian cosmology, the world was created when the first God stood on a mound that emerged from limitless and undifferentiated darkness and water. This represents the time of each year when the land began to reappear from beneath the annual floodwaters. Papyrus swamps were therefore seen as prolifically fertile regions that contained the beginnings of creation. Ceilings in temples and tombs were frequently supported with columns in the form of papyrus plants, turning their architectural settings into models of this primeval swamp. The massive columns are referred to as 'papyriform' with open capitals representing the umbels of the papyrus plant and the columns representing bundles of culms bound together to provide strength. These distinctive features of ancient Egyptian architecture continue to inspire architectural designs today.

The swamps were seen as the marginal zones at the edges of the ordered cosmos, symbols of the untamed chaos that surrounded and perpetually threatened the Egyptian world. The swamps were the setting for ritual hunts as they were seen as teeming with wild birds and fish as well as dangerous animals such as hippos and crocodiles that represented Egypt's enemies. The single-handed defeat of these chaotic creatures by a king or noble, often depicted on the walls of temples and elite tombs, was emblematic of the maintenance of the ordered cosmos against the forces of degradation of the universe.

Papyrus swamps were the location of one of the great mythic cycles central to Egyptian religion, where the goddess Isis took her infant son Horus to the papyrus swamps in the north to conceal him from her brother Seth, who had murdered her husband Osiris and usurped his throne. Here, Horus grew to manhood, hidden among the papyrus culms and umbels whose rustling sounds soothed him and masked his cries, until he emerged to defeat his wicked uncle and reclaim his patrimony. Horus was protected and nursed while a baby by the goddess Hathor, who was worshipped in the ritual of the 'shaking of the papyrus'. This ceremony was performed in the swamp by a celebrant who shook actual culms of papyrus.

Farming was the basis of the entire Egyptian civilisation thanks to the river Nile. The ancient Egyptian calendar was divided into three seasons: *Akhet,* the season of inundation, marked by the flooding of the Nile, *Peret*, the season of growth and emergence, and *Shemu*, the season of harvest. The growth of crops depended on the summer inundation that naturally irrigated the floodplains. Soils enriched with silts deposited by the annual floods were able to support the growth of a wide range of crops including cereals, legumes and fruits for subsistence agriculture. Over time, the Egyptians harnessed and improved the distribution of the flood waters to support an agricultural system that could produce much more food than was required for their subsistence. Large scale cereal production for export eventually developed and this allowed the Pharaohs to build a wealthy empire. Resources became available for managing the activities of large groups that formed communities containing individuals with specialist skills as craftsmen and artists who were responsible for the remarkable artistic and intellectual achievements of ancient Egypt.

The magnitude of the annual cycle of water level of the Nile and its flood plain was of vital importance in determining how well the major food crops for the Egyptians yielded. If the levels were too low then drought could inhibit the growth, but if they were too high they could submerge the crop and waterlogging could cause crop failure. This single factor of flood level became the most important determinant of Egypt's economic activity and tax income. In order to monitor flood levels, nilometers were constructed to measure the level of the water throughout the crop growing areas. These are columns on which a series of carvings are made. If the water level failed to reached the lower mark, then everyone knew that food shortages and starvation would strike in this particular year. If the water reached more than the prescribed 25 marks it would be too high and standing for too long. It would likely therefore destroy the year's crops. The Egyptian state had a special water administration that would monitor the flood heights and try to estimate the peak levels. This allowed predictions about harvest outcome to be made as well as about how to implement measures to attempt to regulate water levels. It also determined how much tax to demand from the farmers.

Papyrus Plant Structure

Papyrus plants have a very unusual structure. When mature it normally spends most of its lifetime floating on water, supported by large buoyant rhizomes with long extensions of roots reaching out for water and nutrients. Rhizomes are modified stems that grow horizontally and at intervals along the stem they produce vertically growing branches, referred to as culms or stems, which are triangular in cross-section and are often up to 5 metres tall (Figure 1.1). The culms are topped by what is botanically a flowering structure called an umbel. The umbel consists of a hundred or more cylindrical shoots, the rays, at the end of which are usually three leaf-like structures which look very much like grass leaves. These structures are called bracteoles. In a mature stand of papyrus the culms are spaced close enough to each other to ensure that the umbels form a completely closed canopy which intercepts virtually all the sunlight incident on it. This

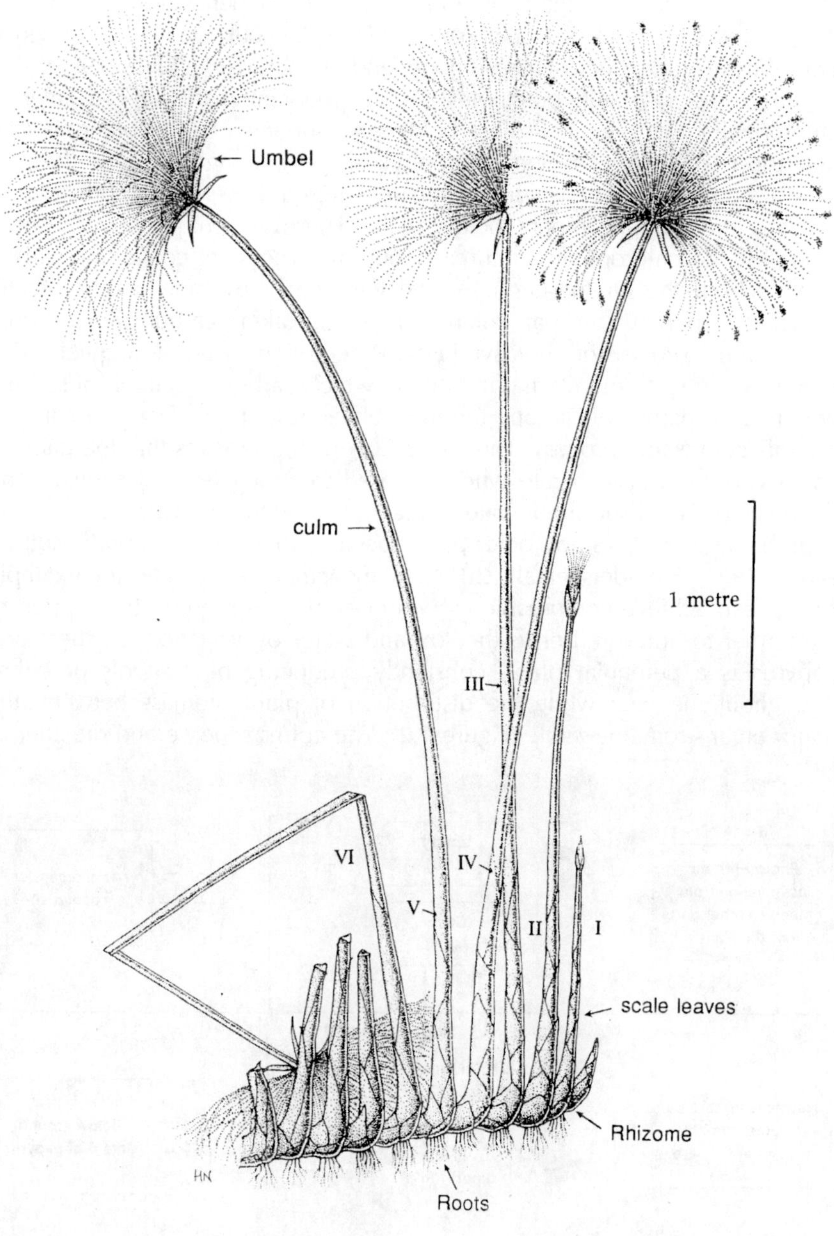

Figure 1.1 A *Cyperus papyrus* plant showing, on a single rhizome, the typical range of culm units of increasing age from I to VI.
Source: Jones and Muthuri (1997)

means that typically less than 10% of the incident light penetrates to the base of the canopy as the rest has been absorbed by the plant chlorophyll and is used to drive the process of photosynthesis (Jones and Muthuri, 1985, Jones, 1987).

The culms, rays and bracteoles of the emergent plant have a layer of chlorophyll-containing cells just below the surface of the epidermis, the outermost layer of cells. The chlorophyll absorbs sunlight to drive the process of photosynthesis to produce sugars which are eventually converted to starch – the building blocks of the plant. There are enzymes in the cells alongside the chlorophyll which carry out this process of converting carbon dioxide from the air to starch. As we will see below, the photosynthetic enzymes in papyrus are particularly efficient at taking carbon dioxide from the air in the process of photosynthesis. When we measure how quickly this process occurs, by monitoring the rate at which carbon dioxide is absorbed, we find that papyrus is among the most efficient group of plants at carrying out this conversion process. This rapid CO_2 fixation means that the papyrus has a very high growth rate when exposed to strong level of sunlight and there is a plentiful supply of water. In fact, its growth rate over a whole year is probably as high as any other plant species (Mnaya et al., 2007; Muthuri et al., 1989; Saunders et al., 2014). Some cultivated plants, for example maize, can achieve comparable growth rates but this requires the support of the farmer to add fertiliser to the soil and water by irrigation. Furthermore, papyrus is a perennial plant, constantly producing new shoots or culms throughout the year while the distribution of plant biomass between the major organs remains stable (Figure 1.2). The culms senesce and die after as

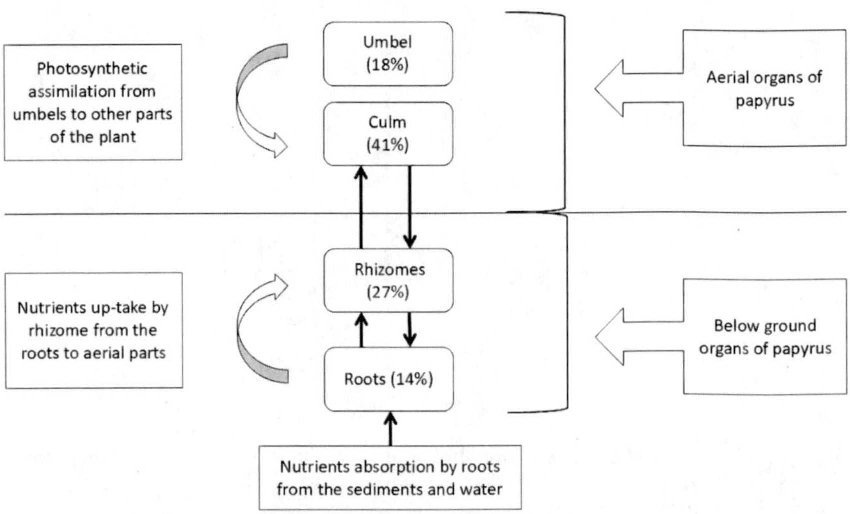

Figure 1.2 Schematic diagram showing percentage distribution of biomass in papyrus plant organs.
Source: Mnaya et al. (2007)

little as six months but they are constantly replaced by new fast-growing culms, so that the plant is constantly rejuvenating itself in tropical and subtropical climates. The dead culms collapse into the water and begin to rapidly decompose under tropical temperatures. However, as they sink below the surface the process of decomposition begins to slow because of the lack of oxygen. Under these anoxic conditions, as the stems sink to the bottom of the water column they build up a thick layer of partly decomposed detritus which is called peat. An area of floating papyrus could potentially have a thick deposit of peat below the rhizomes. However, water movements can frequently lead to this being swept away and being deposited elsewhere or further decomposed where oxygen levels in the water are higher. Under anoxic conditions decomposition may be slower but it does not stop all together as there are decomposing bacteria that are able to function without oxygen. These anaerobes, operating at low oxygen conditions, produce methane rather than CO_2 in the process of respiration. This methane is largely insoluble in water and is released as gas bubbles to the atmosphere. Methane is a potent greenhouse gas and is responsible for some of the global warming which we are currently experiencing. It is estimated that natural wetlands release 20% of global methane emissions, which is 32 times more powerful than CO_2 over a 100-year timeframe (Candry et al., 2023).

Colonising Wetlands by Asexual and Sexual Reproduction

Papyrus colonises the fringes of open water as a floating mat of rhizomes, often tethered to the substrate by roots and then extending out into the open water. Over open water there is no connection with the substrate which means that the floating mat can adjust to changes in the water level. There is often a decline in vigour away from the open water which suggests that its growth is dependent on nutrients carried by water runoff from the land. The rhizomes that occupy open water are usually loosely spreading but they become more tightly clumped as rhizome growth continues, forming a dense mat that is able to support the weight of medium sized animals including the Sitatunga or marsh buck (*Tragelaphus spekii*), a swamp-dwelling antelope found throughout central Africa. Areas of floating mat often break off from the fringe to form floating rafts. These young floating papyrus mats can be dispersed by wind and wave action and may join existing mature stands elsewhere or establish new populations along formerly unoccupied shoreline. These papyrus rafts are referred to as 'sudd', derived from the Arabic meaning 'obstacle' or 'blockage'. Ellery et al. (1995) has described how these floating rafts play a part in the blocking of drainage channels. This is a complex process which starts with the aggradation of the channel bed and accumulation of coarse sediment associated with the decline in the current velocity as the flow of water is restricted. This is followed by encroachment of the channel from the margins of the papyrus mat and by the development of permanent blockages.

The variable nature of annual and longer term patterns of rainfall in the tropics means that tropical wetlands can frequently undergo large-scale fluctuations in water level. Receding water levels, known as a 'drawdown' are followed eventually by a recovery and possibly flooding as water levels rise. For papyrus, extended drawdowns means that the free access to water when vegetation is floating is lost and soil water reserves in the exposed sediment are also progressively lost. The 'stranded' papyrus vegetation can survive for some time by tapping the detritus for water, but growth is quickly severely inhibited (Boar, 2006). Gaudet (1977a) describes the consequences for species dynamics of a natural drawdown on Lake Naivasha in Kenya and the subsequent formation of a papyrus 'fringing swamp' as the water levels rise again (Figure 1.3). In response to these fluctuating conditions many wetland species have evolved characteristics that allow them to live in alternatively wet and dry habitats. When the water levels rise again the fringing papyrus is generated by the germination of papyrus seeds which are buried in the sediment. The seedlings initially compete with hygrophilous annuals but later grow up in the rising water as perennial emergent macrophytes (Gaudet, 1977a). Although the cycles of drying and flooding poses a threat to survival of floating stands that have expanded by vegetative reproduction, the drawdowns offer the opportunity to regenerate by seedling recruitment of germinating seeds (Terer et al., 2014). The process of seed production is the result of sexual reproduction which generates

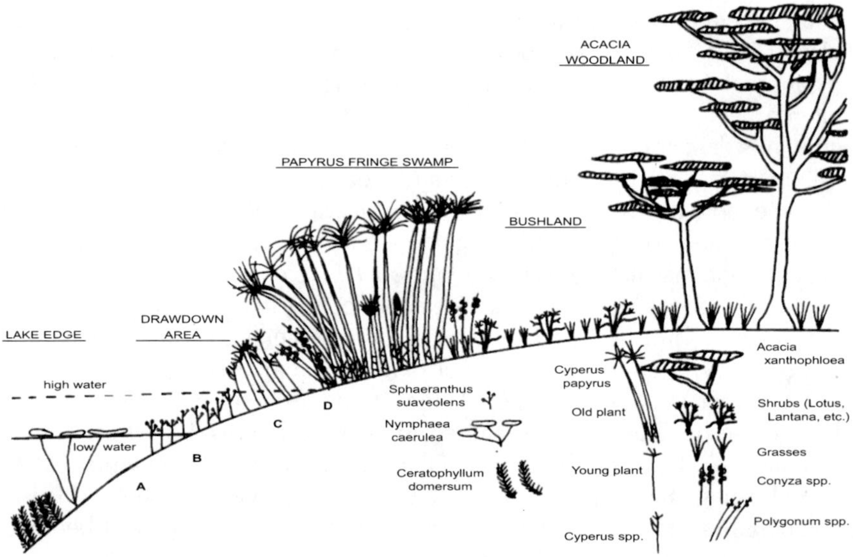

Figure 1.3 The lake-side communities and their zonation, showing location of papyrus fringing swamp on Lake Naivasha, Kenya.
Source: Gaudet (1977a)

genetic diversity and offers the opportunity to adapt to changing conditions. By combining two strategies – seedling development during drawdown and vegetative reproduction year round – papyrus has adapted to the drawdown/flooding conditions in the tropics (Terer et al., 2015).

Following germination and seedling growth the rhizome that develops is referred to as a ramet and the mass of vegetative growth that develops from a collection of ramets is called a genet. Each ramet comprises a portion of rhizome, with a basal sheath of leaves that protect the terminal and axillary buds of the rhizome. The axillary buds grow to form the triangular (three-sided) culms in cross section, which supports the umbel. The umbel is the primary photosynthetic organ but it also bears the inflorescence when it is produced. The newly produced culm bases are covered by papery scales, which are actually reduced leaves. These are triangular in shape and a reddish-brown colour. The nondescript, greenish-brown flowers that are produced at the ends of the rays are wind pollinated. Eventually, brown nut-like fruits are produced and the mature fruits are dispersed in water after they fall from the plant.

Flowering in papyrus appears to be stimulated by stress conditions, particularly drought, and it is generally found that in actively growing stands few of the umbels mature to produce seeds. Its adaptation to cycles of drying and flooding is enabled by the production of large numbers of seeds that can lie dormant in exposed sediments for many years of drying and then showing unlimited clonal propagation when the plants are floating on recovery of water levels. Vegetative reproduction therefore plays the main role in the maintenance of local populations but sexual production of seeds and their dispersal is more effective for colonising new sites. When stimulated to flower the papyrus umbel has a prolific seed producing potential. When five representative umbels were sampled by Terer et al. (2014) they were found to produce between 98,000 and 327,000 seeds each and as many as one third of the papyrus umbels in the canopy were flowering at any given time. Flowering papyrus therefore produces an enormous seed rain onto the swamp and lake shore. The germination rate of seeds under suitable conditions are also high at 25% after three weeks. Boar (2006) found seeds germinated and seedlings survived best when the wetland sediment was saturated but not flooded. Papyrus seeds germinated quickly in rewetted sediment as water levels rise but new swamps only form on shallow gradients or when the water level rises slowly enough for seedlings vertical growth to keep pace with water level. In the developing papyrus seedlings the first developing true leaves are the dominant photosynthetic structure and the roots penetrate the substrate. The true leaves that appear during seedling emergence and extension of the culms, senesce under the mature canopy formed of culms and umbels. The rhizomes develop and extend as the plant matures and they float on the rising water where they eventually form floating mats.

In recent years it has been shown by Triest et al. (2014) that the genetic diversity of papyrus can vary, depending on the history of the swamp. In areas of long-lived vegetation, where the plant has survived and spread by clonal propagation, diversity is low. This may mean that some plants associated with permanent lakes and rivers are possibly genetically identical to those that were present hundreds of years ago. Triest et al. (2014) used molecular techniques to estimate the clonal diversity of papyrus bordering Lake Naivasha in Kenya, which has been severely disturbed by anthropogenic activity and has been subjected to a series of drawdowns. It was shown that the swamp consists of a series of genetically diverse clones which originated through seedling recruitment from seeds that had been dispersed during previous drawdowns. Triest et al. (2014) suggested that this diversity at local scale should allow rapid adaptive responses and strengthen the ecological stability of papyrus populations subjected to environmental change. It was concluded that in papyrus, seedling recruitment provides for the replenishment and maintenance of the genetic diversity necessary for adaptation to changes in environmental conditions.

For wetland plants that have evolved to live in alternatively wet and dry habitats, drawdown can be a very positive event that promotes regeneration by seedling recruitment. By employing both sexual and unlimited clonal propagation, papyrus has adapted to the normal climatic cycles of drying and flooding. Mixed reproductive strategies combined with ample gene flow across relatively small spatial scales appear to be responsible for maintaining a high level of genetic diversity (Lusasi, 2017). Maintenance of high genetic diversity and high gene flow through sexual reproduction is important for the long term success of the species under changing climatic conditions. In particular, after extended drought the existence of a substantial seed bank in the swamp sediment allows the recuperation of the plant population after subsequent rewetting. Our improved understanding of the connection between drawdown and population dynamics will be crucial to putting in place conservation measures to reverse population loss.

Survival in Wetlands – Floatation and Internal Ventilation

The survival of papyrus in wetland conditions is enabled by the development of some important internal structural characteristics. The internal anatomy of the most prominent aerial vegetative tissues of papyrus can be discerned by examining cross sections of the bracteoles and culms. The outermost layer is the epidermis. This consists of a single layer of thick walled cells which in the culms and bracteoles are interspersed with pairs of guard cells which control the opening of the stomatal pores. Internal to the epidermis are one or two layers of supporting sclerenchyma cells. These are thick walled cells which are dead once mature. Below the sclerenchyma are several layers of photosynthetic tissue containing chloroplasts in thin walled parenchyma cells. Evenly distributed around the culm and embedded in the photosynthetic tissue is a ring of small vascular bundles, consisting of xylem and phloem which allow the movement of water upwards in the transpiration stream, and sugars produced by photosynthesis in the distinctive bundle sheath cells that surround the vascular bundles. These bundle sheath cells are tightly packed with chloroplasts. Internal to the outer vascular bundles are larger but more widely spaced vascular bundles that become less frequent and eventually absent from the centre of the culm in cross section. Within the rings of vascular bundles the most characteristic anatomical feature is the development of aerenchyma: large intercellular spaces which form regularly arranged chambers throughout their vegetative tissues. The aerenchyma is formed of thin-walled parenchyma cells which are stacked in a honeycomb like network creating vertical intercellular spaces which interconnect throughout the culms and rhizomes. An important adaptive advantage that this gives is to make the vegetative structure of buoyant rhizomes the ability to float on water to form the rhizome mat. The thick mats of papyrus rhizomes are able to move up and down with the fluctuations in water level, which ensures that the canopy is not inundated by the water. The development of aerenchyma is also an important adaptation that allows rapid diffusion of oxygen from the air to the roots and rhizomes located in the anaerobic substrate of mud or peat developing from decomposing vegetation, and thus allowing aerobic respiration to continue in an anaerobic environment. In papyrus there appears to be a continuous system of air spaces extending from the sub-stomatal cavities to the root, interrupted by occasional diaphragms of a few cells thickness (Li and Jones, 1995). It has also been demonstrated that oxygen can diffuse out of the roots of papyrus with the resultant oxidation of reduce soluble elements such as reduced Iron and manganese (Brix et al., 1992). A consequence of this process can be a thick deposit of insoluble hydrated iron oxides coating the root system. As the reduced forms of iron and manganese are potentially toxic to plants when, in their soluble state, they are carried into the plant in the transpiration stream their oxidation and deposition at the root surface effectively detoxifies them.

The movement of oxygen through the aerenchyma of wetland plants can occur by simple diffusion but there have been several demonstrations of it being enhanced by a process of internal pressurisation and convective through-flow (Brix et al., 1992). By this process, mass flow of air occurs at faster rates than simple diffusion, resulting in the maintenance of higher oxygen concentration in the rhizomes. This movement is ascribed to a 'pressurised pump' driven by the gradients in temperature and water vapour between the internal gas spaces of the plant and the ambient atmosphere. The pressurised pump creates an airflow down the plant which in the case of papyrus may be relieved by a reverse flow back up the culm carrying CO_2 generated in rhizome respiration to the aerial parts of the plant where the CO_2 can be refixed in photosynthesis (Singer et al., 1994; Li and Jones, 1995). This bidirectional gas flow has not been demonstrated in papyrus but it has been shown to exist in *Nelumbo nucifera* (lotus) (Mevi-Schutz and Grosse, 1988). In darkness, the stomata on the epidermis of the culm close and the carbon dioxide concentration in the culms has been shown to increase significantly above that of the ambient air. In the light the CO_2 concentration falls as it is recycled via photosynthesis (Li and Jones, 1995). Results suggest that 35–57% of the carbon dioxide respired by the culm pith and rhizomes can be re-fixed by culm photosynthesis. In the daytime the culm is also fixing CO_2 from the air diffusing into the culm through stomatal pores.

Rapid Photosynthesis and the Origins of C_4 Photosynthetic Pathway in Papyrus

A characteristic of papyrus that may contribute to its success in becoming the dominant species in vast areas of wetlands is its rapid growth rate which means that it can out-compete other species for space and light. In many wetlands papyrus is virtually a monoculture. Measurements of the productivity of papyrus have shown that in many localities it has one of the highest recorded plant growth rates (Saunders et al., 2007). Growth of all green plants is the result of photosynthesis where atmospheric CO_2 is assimilated to form the plant biomass. Photosynthetic CO_2 fixation has provided the carbon for life on Earth for at least the past 2.7 billion years (Osborne and Beerling, 2006). More than 99% of the history of this process has been dominated by a form of assimilation called 'C_3 photosynthesis' because the first products are carboxylic acids formed of three linked carbon atoms. However, between 25 million and 32 million years ago a revolutionary innovation evolved in tropical plants in the form of a solar-powered CO_2 pump based on four-carbon acids which could boost photosynthesis in hot conditions. It was called C_4 photosynthesis because the first products of the process were four-carbon acids. Theoretically, C_4 photosynthesis has, in warm climates, a 40% greater ability to covert sunshine into biomass than C_3 species and maximum growth rates of the most productive C_4 species are 40–50% greater than those of the most productive C_3 species. Papyrus, in common with many of the tropical and sub-tropical *Cyperaceae*, uses C_4 photosynthesis (Jones and Milburn, 1978; Jones, 1987).

The major evolutionary development of C_4 photosynthesis now accounts for 20–30% of terrestrial CO_2 fixation on Earth. When the biochemical pathway for the conversion of carbon dioxide to sugars used in growth of plants was first elucidated, it was shown that the conversion was catalysed by an enzyme abbreviated as rubisco (ribulose-1,5-bisphosphate carboxylase/oxygenase). Initially it was believed that in all photosynthesising plants the first product produced when CO_2 was 'fixed' was the three carbon compound phosphoglyceric acid, and consequently the process was called C_3 photosynthesis. Subsequently it was shown that in some plant species there was a variant of this pathway where the first product of CO_2 fixation is a four carbon organic acid, rather than a three carbon compound. This reaction is catalysed by an enzyme called PEP-carboxylase and the process was referred to as the C_4 pathway or C_4 photosynthesis. The important characteristic of PEP-carboxylase is that it has a much higher affinity for CO_2 than rubisco. It was also demonstrated that plants with C_4 photosynthesis, particularly under the climatic conditions of the tropics and sub-tropics, has higher rates of photosynthesis in strong sunlight and can grow faster and outcompete plants with C_3 photosynthesis (Beerling, 2017). Osborne and Beerling (2006) have suggested that the C_4 mechanism arose as a 'fix' for the intrinsic inefficiency in the enzyme rubisco. It is known that this enzyme evolved when the Earth's atmosphere contained CO_2 at concentrations as much as 100 times today's level and there were very low concentrations of oxygen. Under these conditions, the fact that rubisco had a relatively low affinity for CO_2 as well as an oxygenase activity was insignificant. However, as atmospheric CO_2 conditions declined over time and oxygen concentrations increased it became more difficult for the carboxylase reaction to assimilate CO_2 and the oxygenase activity increased causing a net loss of CO_2 via what is referred to as the photorespiration pathway. It was also demonstrated that C_4 photosynthesis involves a series of biochemical and structural adjustments which allows C_4 plants to exploit PEP-carboxylase to efficiently fix CO_2 from the atmosphere (Sage, 2004, 2016). The initial fixation of CO_2 by PEP-carboxylase to yield the four-carbon compound oxaloacetic acid (OAA) is the way of capturing more CO_2 from the atmosphere and is followed by the movement of the resulting four-carbon acids to an interior compartment where rubisco is localised. Here CO_2 is released by decarboxylation and this leads to a rise in CO_2 concentration to a level that nearly saturates the rubisco active site. Three decarboxylation enzymes have been identified and their presence is the basis for naming three sub-types of C_4 photosynthesis. The three decarboxylation enzymes are NADP-malic enzyme found in papyrus, NAD-malic enzyme, and PEP carboxykinase.

Leaves of species with C_4 photosynthesis also have lower nitrogen contents than those with C_3 photosynthesis because they need less of the enzyme rubisco and its nitrogen content to assimilate a given amount of carbon. C_4 species are therefore more efficient in their use of nitrogen. This confers a particular advantage in environments where nitrogen availability is low (Gaudet, 1977b). Freshwater wetlands are often nutrient, and in particular nitrogen, limited. Under these circumstances a C_4 species like papyrus might be expected to achieve a higher productivity in freshwater wetlands and therefore be more competitive than species with C_3 photosynthesis (Jones, 1987).

Surprisingly, we now know that the C_4 pathway has evolved independently over 45 times and is found in 19 families of angiosperms. This represents one of the most convergent of evolutionary steps in plants. C_4 photosynthesis is now present in about 7,500 species of flowering plants, about 3% of the estimated 250,000 land plants. However, their competitive advantage is reflected in the fact that they occupy about a fifth of the global vegetated land surface and account for up to 30% of the primary productivity of the terrestrial biosphere. Perhaps, even more surprisingly, the dominance of the tropical grassland and savannah environments was only achieved in the last 10 million years, which is less than 3% of the time that terrestrial plants have existed on Earth (Beerling, 2017).

The anatomical changes required for C_4 photosynthesis involve the modification of the C_3 leaf structure to form the inner compartments where the rubisco enzyme is localised and CO_2 can be concentrated. This normally involves the formation of a wreath-like cell arrangement termed 'Kranz anatomy'; the name coming for the German word for wreath (Brown, 1975). The inner layer holding rubisco in large chloroplasts is arranged around the vascular bundles in the leaf, to ensure rapid access to the phloem that transports the products of photosynthesis. PEP-carboxylase is located in the outer, mesophyll layer. In total, 15 distinct types of Kranz-type leaf anatomy have been identified in the various dicot and monocot lineages of C_4 photosynthesis. In all cases the outer wall of the Kranz structure has been modified to reduce the rate of CO_2 leakage and therefore traps CO_2 inside (Carolin et al., 1977).

Questions about why, how and when C_4 photosynthesis in plants evolved have intrigued researchers for several decades. The picture is still not clear but attempts to provide answers have provided a much more detailed understanding of the process itself (Sage, 2004). The story revolves around the fact that rubisco is the only enzyme that catalyses the net fixation of CO_2 into organic molecules and, as we have seen, this enzyme evolved early in the history of life when atmospheric CO_2 concentrations were probably much higher than they are now. It is important to know that all C_4 plants operate a complete C_3 cycle so the C_4 pathway supplements by a series of irreversible steps, rather than replaces, C_3 photosynthesis. Also, in photorespiration, rubisco oxidises CO_2 to phosphoglycolate which is metabolically useless and is toxic if accumulated in plant cells. At current concentrations of CO_2 and O_2 photorespiration reduces photosynthesis by over 30%, particularly at warmer temperatures (> 30°C). During the earth's early history photorespiration was negligible due to elevated CO_2 and low O_2 levels in the atmosphere and only during the carboniferous period (280–340 million years ago) and in the past 35 million years have atmospheric conditions favoured significant levels of photorespiration (Osborne and Beerling, 2006). However by suppressing photorespiration the C_4 pathway is described as supercharging photosynthesis in C_4 species (Sage, 2016). For a variety of reasons evolution failed to modify rubisco to reduce the oxygenase activity but at the same time all the enzymes for increasing CO_2 concentration at the site of carboxylation were already present in plants with C_3 photosynthesis, serving a variety of functions including stomatal functioning and accumulation of C_4 acids in succulent plants. These have been added to existing C_3 mechanisms in a modest series of incremental steps during evolution (Christin and Osborne, 2014).

Assessing the presence of C_4 photosynthesis in geological time is determined primarily by screening for C_4 isotope signals in geological deposits, including fossilised animals that have consumed plant material (Osborne and

Beerling, 2006). The carbon in CO_2 occurs as two stable isotopes and it has been established that C_3 and C_4 plants differ in the ratio of these isotopes that they fix in photosynthesis. Furthermore, anything derived from the fixed carbon should be reflected in the carbon isotope ratio of the source plant. Carbon isotope ratios of fossil soils, animal tooth enamel, and fossil eggshells show a dramatic shift from C_3-like to C_4-like values between 8 and 5 million years ago in Africa, South America, China and North America, suggesting that there was a widespread expansion of C_4 dominated biomes during this period. However, shifts in isotopic ratios from soils and herbivores in East Africa which were earlier, at 14–20 million years ago, have also been reported. C_4 isotope signals in geological deposits have been useful in indicating when and where C_4 plants were common on the landscape but they are less useful in indicating when C_4 photosynthesis first evolved as C_4 plant species may have initially remained minor components of ecosystems.

An alternative approach to assess the earliest origins of C_4 photosynthesis, used by molecular biologists, is to compare gene sequences in existing taxa to estimate the divergence time of C_4 lineages (Christin and Osborne, 2014). This analysis has suggested that the C_4 pathway evolved in grasses about 30 million years ago, probably in response to the decline of atmospheric CO_2 concentration. However, C_4 Cyperaceae, including papyrus, appear to have evolved considerably later. Here the first C_4 evolution has likely occurred between 19.6 and 10.1 million years ago but most families arose during the last 12 million years (Besnard et al., 2009). The drivers of these of these evolutionary events may have been influenced by the Oligocene (39–24 million years ago) CO_2 decline, but other factors such as heat, drought, fire and other disturbances may have been locally selective for the C_4 pathway (Osborne and Freckleton, 2009). There is evidence that C_4 Cyperaceae generally occupy wetter habitats than C_4 grasses within fire-adapted vegetation and they probably evolved their C_4 systems in a wetland context (Besnard et al., 2009) and here the nitrogen limited status of wetlands may have favoured selection for the higher nitrogen use efficiency of the C_4 syndrome.

Nutrient Dynamics of Papyrus Swamps

As we have seen, swamps are continually inundated with water to variable depths, although they can experience irregular but important drops in water level known as drawdowns when sediments are exposed and dry out. In the transition from land to water, the shallow water swamps are characterised by herbaceous plants such as *Typha* spp. and *Phragmites* spp., with erect shoots that rise out of the water to a height of up to four metres while they are firmly rooted in the sediment. In deeper water, papyrus forms the dominant swamp vegetation which is rooted in shallower water but then in deeper water floats, supported by a thick mat of buoyant rhizomes. These two types of swamp have different nutrient pathways. In the case of the rooted plants, nutrients are first incorporated within the sediments and uptake by the plants occurs through extensive root systems. In the case of the papyrus floating mat, dissolved nutrients are efficiently absorbed by the plant roots directly from the water column. Minerals in the wetland vegetation are immobilised in the swamp vegetation by incorporation into the plant biomass. Once immobilised by the vegetation they may be recycled several times between the aerial organs and rhizomes, but eventually they are released to the water flooding the mat and ultimately to the detritus below the mat (Thompson, 1976).

In the growth cycle of papyrus, culms are produced continuously and they extend up to the top of the canopy carrying an umbel that expands to form leaf-like bracteoles that capture light and carry out photosynthesis. The culms are closely spaced to form a closed canopy that intercepts as much as 95% of the incident photosynthetically active radiation before it reaches the water or rhizome mat (Jones and Muthuri, 1985). The primary production of papyrus as a result of photosynthesis has been estimated by measuring the accumulated growth of individual culms as well as by measuring canopy photosynthesis (Muthuri et al., 1989; Saunders et al., 2007). These measurements have demonstrated that the growth rates of the canopy are some of the highest ever recorded (Saunders et al. 2014). The umbels remain active for several months but then the culms collapse into the canopy as they begin to senesce and eventually die. Dead and decomposing plant material breaks away and accumulates as peat beneath the open mat. Soluble nutrients then re-enter the water column before they are taken up again by the roots. Loss of nutrients can occur through the export of detritus from the swamp and is largely dependent on the rate of water through-flow in the swamp. The detritus is utilised mainly in the open lakes or down-stream as a food source and physical substrate for microorganisms.

Although papyrus swamps have high production rates they do not support correspondingly high herbivore production, probably because of the low nutritive value of the vegetation. Among the few large herbivores adapted to swamp conditions are the Nile lechwe (*Kobus megaceros*) and situtanga (*Tragelaphus spekii*), which have greatly elongated hooves for walking across unstable substrates. Very little of the net production in swamps is therefore grazed, and losses of nutrients as offtake by grazers is correspondingly small.

Swamp waters supporting dense vegetation have low levels of dissolved oxygen, they are anaerobic, which influences the sediment chemistry and the organisms that live there (Schlesinger, 1997). In particular, deoxygenated waters favour denitrifying bacteria which lead to gaseous nitrogen loss. Denitrifying bacteria use oxidised nitrogen compounds, such as nitrate and nitrite in the absence of oxygen as a terminal electron acceptor. In a mature wetland swamp the standing biomass remains stable and the primary production replaces dying and senescing plant material. Under aerobic conditions this vegetation decomposes rapidly but under anaerobic conditions below the surface of the water the process of decomposition slows markedly and this leads to the accumulation of organic detritus (peat).

The availability of nutrients from water inflow and sediments in wetlands varies greatly from oligotrophic (low nutrient content) to eutrophic (nutrient rich). Nutrient supply may be seasonal with higher inputs entering with wet season inflows and in more densely populated regions there is nutrient enrichment from agricultural activity The vegetation in wetlands is a sink for nutrients, giving them the ability to 'clean up' eutrophic waters flowing through them (Kansiime et al., 2007). Uptake of nutrients is related directly to the growth rate of the vegetation with the plants absorbing the largest quantities from the water and substrate during their exponential growth phase. In floating papyrus wetlands the continuous movement of water means that the vegetation is growing in a flowing nutrient solution, where nutrient uptake is particularly efficient. Once taken up by the roots, the nutrients are distributed throughout the plant during growth but are subsequently released and recycled during senescence.

Papyrus growth traps large amounts of available nutrients, especially nitrogen and phosphorus, in actively growing vegetation (Gaudet, 1979; Kyambadde et al., 2004; Kelderman et al., 2007). The supply of nitrogen to papyrus is a complex balance between the processes of nitrogen fixation and denitrification. Calculations of the nutrient balance of pristine papyrus swamps, low in nutrients and uncontaminated by runoff from surrounding agricultural land, show that the major portion of nitrogen in the swamp must come from nitrogen fixation as the amounts of nitrogen imported are insignificant in terms of the total amount of annual uptake by the papyrus. In the dense tangle of roots in the papyrus mat there is intense localised nitrogen fixation by free living bacteria, which supplies nitrogen to the papyrus plant (Mwaura and Widdowson, 1992). A large proportion of the nitrogen in plants is allocated to photosynthetic enzymes and the rate of leaf photosynthesis is linearly related to nitrogen content in papyrus (Jones, 1986). A physiological advantage of C_4 photosynthesis is that it utilises nitrogen more efficiently than C_3 photosynthesis, so that C_4 plants can achieve higher rates of leaf photosynthesis at given plant nitrogen concentrations.

Nutrient retention is influenced by natural and anthropogenic disturbances. Natural disturbance is largely influenced by flooding and drying cycles and anthropogenic activities are largely influence by harvesting and permanent conversion to agricultural activities. Hes et al. (2021) have shown how seasonal flooding and harvesting of papyrus influence the retention of nitrogen and phosphorus by the swamp. They showed that nutrient retention was higher in permanently inundated zones than seasonally inundated zones, and higher in sites with low disturbance than harvested sites. The greater retention of nitrogen and phosphorus were the result of development of large mature culms of papyrus under permanently inundated and undisturbed conditions. Models of the role of papyrus vegetation in nutrient cycling have been developed by van Dam et al. (2007) and Hess et al. (2014) for floating and rooted papyrus respectively. These models are based on a description of the processes of photosynthesis and growth, nutrient uptake and decomposition and can be used to understand the processes contributing to nutrient retention, and in particular evaluate the effects of potential management strategies such as regular harvesting, on the nutrient absorption capacity of the papyrus wetlands (Hes and VanDam, 2019).

Conclusions

Papyrus (*Cyperus papyrus*) is the largest sedge in the world and one of the largest entirely herbaceous plants. It normally attains a height of 3–5 m, but can reach 10 m at higher altitude sites. It occurs in nearly monoculture stands, creating extensive areas of wetlands in river valleys and near lake edges, most predominantly in eastern, central and southern Africa. The most extensive papyrus swamps are associated with large shallow lake systems and river valleys with slow response times during the wet seasons. It can also form floating mats up to 1.5 m thick which are largely independent of water depth. The aerial vegetation consists of closely spaced culms which are topped by an umbel consisting of hundreds of cylindrical rays and flattened, leaf-like bracteoles which form the main photosynthetic surface of the plant as true leaves are absent. The total land area covered by papyrus wetlands is not known for certain, and it also varies seasonally with rainfall, but one estimate puts it as high as 40,000 km^2 (Hughes and Hughes, 1992). The importance of papyrus is not only as a result of its vast aerial coverage but because it provides vital ecosystem functions and services for millions of people, particularly in the River Nile catchment.

Papyrus is a plant species that has evolved to exploit an environment that is almost permanently flooded in a tropical climate. These conditions are optimal for growth of papyrus and as a result it is among the fastest growing plant species in the world. It has a number of structural, morphological and ecological characteristics that have the potential to contribute to its success in forming extensive areas of papyrus monoculture. Structurally, the papyrus canopy intercepts more than 95% of incident solar radiation and physiologically it has a form of photosynthesis, referred to as C_4 photosynthesis, that has a 40% greater ability to covert sunshine into biomass than the more common species with C_3 photosynthesis, and maximum growth rates that are 40–50% greater than those of the most productive C_3 species.

A distinctive morphological feature present in papyrus, and commonly occurring in wetland adapted plants, are large intercellular and interconnected spaces dispersed throughout the culms and rhizomes. These aid the floatation of the papyrus mat but also enable the diffusion of gases up and down the culm. Concentration gradients drive the movement of oxygen down the culms towards the oxygen-depleted rhizomes and carbon dioxide produced in respiration in the rhizomes moves up the culms where it is assimilated in photosynthesis by chloroplasts in vascular bundles around the perimeter of the culms.

Nutrient capture for growth of papyrus is particularly effective as the roots have access to soluble nutrients that have dissolved in the water that flows from the land and through the swamp. At the same time more nitrogen is made available by microbial associations with nitrogen fixing bacteria in the rhizome mats. An additional advantage provided by the use of C_4 photosynthesis is the higher nitrogen use efficiency of assimilation than in plants with C_3 photosynthesis, which is the consequence of lower investment in nitrogen containing photosynthetic enzymes.

Survival of papyrus in the wetland environment is dependent on the ability for adjustments in its life cycle in response to fluctuations in water level. Variable seasonal and long-term patterns of rainfall means that papyrus wetlands can undergo large-scale fluctuations in water level. Variations in water level have implications for reproduction of the papyrus. Vegetative reproduction plays the main role in maintenance and expansion of swamps when water levels are high but recovery from drawdowns depends on the germination of seeds produce in sexual reproduction. Seeds are produced by mature umbels and deposited on the exposed sediment. Seed production is most effective in colonising new sites.

References

Adam, E. and Mutanga, O. (2009). Spectral discrimination of papyrus vegetation (*Cyperus papyrus* L.) in swamp wetlands using field spectrometry. *ISPRS Journal of Photogrammetry and Remote Sensing*, 64, 612–620.

Beerling, D. (2017). *The emerald planet*. Oxford: Oxford University Press.

Besnard, G., Muasya, A.M., Russier, F., Roalson, E.H., Salamin, N., and Christin, P.A. (2009). Phylogenomics of C_4 photosynthesis in sedges (Cyperaceae): Multiple appearances and genetic convergence. *Molecular Biology and Evolution*, 26, 1909–1919.

Boar, R.R. (2006). Responses of a fringing *Cyperus papyrus* L. swamp to changes in water level. *Aquatic Botany*, 84, 85–92.

Boar, R.R., Harper, D.M. and Adams, C.S. (1999). Biomass allocation in *Cyperus papyrus* in a tropical wetland, Lake Naivasha, Kenya. *Biotropica*, 31, 411–421.

Brix, H., Sorrell, B.K. and Orr, P.T. (1992). Internal pressurization and convective gas flow in some emergent freshwater macrophytes. *Limnology and Oceanography*, 37(7), 1420–1433.

Brown, W.V. (1975). Variations in anatomy, associations, and origins of Kranz tissue. *American Journal of Botany*, 62(4), 395–402.

Candry, P., Abrahamson, B., Stahl, D.A. and Winkler, M.K.H. (2023). Microbially mediated climate feedbacks from wetland ecosystems. *Global Change Biology*, 29, 5169–5183.

Carolin, R.C., Jacobs, S.W. and Vesk, M. (1977). The ultrastructure of Kranz cells in the family Cyperaceae. *Botanical Gazette*, 138(4), 413–419.

Carter, G. S. (1955). *The papyrus swamps of Uganda*. Cambridge: W. Heffer & Sons.

Chan, E.K., Timmermann, A., Baldi, B.F., Moore, A.E., Lyons, R.J., Lee, S.S., Kalsbeek, A.M., Petersen, D.C., Rautenbach, H., Förtsch, H.E. and Bornman, M.R. (2019).

Human origins in a southern African palaeo-wetland and first migrations. *Nature*, 575(7781), 185–189.

Christin, P.A. and Osborne, C.P. (2014). The evolutionary ecology of C_4 plants. *New Phytologist*, 204(4), 765–781.

Dargie, G.C., Lewis, S.L., Lawson, I.T., Mitchard, E.T., Page, S.E., Bocko, Y.E. and Ifo, S.A. (2017). Age, extent and carbon storage of the central Congo Basin peatland complex. *Nature*, 542, 86–90.

Ellery, W.N., Ellery, K., Rogers, K.H. and McCarthy, T.S. (1995). The role of *Cyperus papyrus* L. in channel blockage and abandonment in the northeastern Okavango Delta, Botswana. *African Journal of Ecology*, 33(1), 25–49.

Gaudet, J.J. (1977a). Natural drawdown on Lake Naivasha, Kenya, and the formation of papyrus swamps. *Aquatic Botany*, 3, 1–47.

Gaudet, J.J. (1977b). Uptake, accumulation, and loss of nutrients by papyrus in tropical swamps. *Ecology*, 58, 415–422.

Gaudet, J.J. (1979). Seasonal changes in nutrients in a tropical swamp: North Swamp, Lake Naivasha, Kenya. *Journal of Ecology*, 67, 953–981.

Gaudet, J. (2014). *Papyrus: The plant that changed the world – from ancient Egypt to today's water wars*. New York: Pegasus Books.

Gumbricht, T., Roman-Cuesta, R.M., Verchot, L., Herold, M., Wittmann, F., Householder, E., Herold, N. and Murdiyarso, D. (2017). An expert system model for mapping tropical wetlands and peatlands reveals South America as the largest contributor. *Global Change Biology*, 23(9), 3581–3599.

Hes, E.M.A., Niu, R. and van Dam, A. (2014). A simulation model for nitrogen cycling in natural rooted papyrus wetlands in East Africa. *Wetlands Ecology and Management*, 22, 157–176.

Hes, E.M., and van Dam, A.A. (2019). Modelling nitrogen and phosphorus cycling and retention in *Cyperus papyrus* dominated natural wetlands. *Environmental Modelling & Software*, 122, 104531. doi:10.1016/j.envsoft.2019.104531.

Hes, E.M., Yatoi, R., Laisser, S.L., Feyissa, A.K., Irvine, K., Kipkemboi, J., and van Dam, A.A. (2021). The effect of seasonal flooding and livelihood activities on retention of nitrogen and phosphorus in *Cyperus papyrus* wetlands, the role of aboveground biomass. *Hydrobiologia*, 848, 4135–4152.

Howard-Williams, C. (1977). Swamp ecosystems. *The Malayan Nature Journal*, 31, 113–125.

Howard-Williams, C. and Gaudet, J. (1985). The structure and functioning of African swamps. In P. Denny (ed.), *The ecology and management of African wetland vegetation*. Dordrecht: Dr W. Junk Publishers.

Hughes, R.H. and Hughes, J.S. (1992). *A directory of African wetlands*. Gland: IUCN.

Jones, M.B. (1986). Wetlands. In N.R. Baker & S.P. Long (eds), *Photosynthesis in contrasting environments*. Amsterdam: Elsevier.

Jones, M.B. and Milburn, T.R. (1978). Photosynthesis in papyrus (*Cyperus papyrus*, L.). *Photosynthetica*, 12, 197–199.

Jones, M.B. and Muthuri, F.M. (1997). Standing biomass and carbon distribution in a papyrus (*Cyperus papyrus* L.) swamp on Lake Naivasha, Kenya. *Journal of Tropical Ecology*, 13, 347–356.

Jones, M.B. and Muthuri, F.M. (1985). The canopy structure and microclimate of papyrus (*Cyperus papyrus*) swamps. *Journal of Ecology*, 73, 481–491.

Jones, M.B. (1987). The photosynthetic characteristics of papyrus in a tropical swamp. *Oecologia*, 71, 355–359.

Kansiime, F., Kateyo, E., Oryem-Origa, H., and Mucunguzi, P. (2007). Nutrient status and retention in pristine and disturbed wetlands in Uganda: Management implications. *Wetlands Ecology and Management*, 15, 453–467.

Kelderman, P., Kansiime, F., Tola, M.A. and van Dam, A. (2007). The role of sediments for phosphorus retention in the Kirinya wetland (Uganda). *Wetlands Ecology and Management*, 15, 481–488.

Kyambadde, J., Kansiime, F., Gumaelius, L. and Dalhammar, G. (2004). A comparative study of *Cyperus papyrus* and *Miscanthidium violaceum*-based constructed wetlands for wastewater treatment in a tropical climate. *Water Research*, 38, 475–485.

Li, M. and Jones, M.B. (1995). CO_2 and O_2 transport in the aerenchyma of *Cyprus papyrus* L. *Aquatic Botany*, 52, 93–106.

Li, A., Song, K., Chen, S., Mu, Y., Xu, Z. and Zeng, Q. (2022). Mapping African wetlands for 2020 using multiple spectral, geo-ecological features and Google Earth Engine. *ISPRS Journal of Photogrammetry and Remote Sensing*, 193, 252–268.

Lind, E.M. and Visser, S.A. (1962). A study of a swamp at the north end of Lake Victoria. *The Journal of Ecology*, 50(3), 599–613.

Lusasi, J. (2017). *Cyperus papyrus* in Lake Victoria: Genetic information, utilisation and resource sustainability. *Journal of the Geographical Association of Tanzania*, 36, 73–88.

Mevi-Schutz, J. and Grosse, W. (1988). A two-way gas transport system in Nelumbo nucifera. *Plant, Cell & Environment*, 11(1), 27–34.

Migahid, A.M. (1947). An ecological study of the 'Sudd' swamps of the Upper Nile. *Proceedings of the Egyptian Academy of Sciences*, 3, 57–86.

Migahid, A.M. (1948). *Report on a Botanical Excursion to the Sudd Region*. Cairo: Fouad University Press.

Mitch, W.J. and Gosselink, J. G. (2007). *Wetlands*, 4th edition. Chichester: Wiley.

Mnaya, B., Asaeda, T., Kiwango, Y. and Ayubu, E. (2007). Primary production in papyrus *(Cyperus papyrus* L.) of Rubondo Island, Lake Victoria, Tanzania. *Wetlands Ecology and Management*, 15, 269–275.

Muthuri, F.M., Jones, M.B. and Imbamba, S.K. (1989). Primary productivity of papyrus (*Cyperus papyrus*) in a tropical swamp; Lake Naivasha, Kenya. *Biomass*, 18, 1–14.

Mwaura, F.B. and Widdowson, D. (1992). Nitrogenase activity in the papyrus swamps of Lake Naivasha, Kenya. *Hydrobiologia*, 232, 23–30.

Osborne, C.P. and Beerling, D.J. (2006). Nature's green revolution: The remarkable evolutionary rise of C_4 plants. *Philosophical Transactions of the Royal Society B: Biological Sciences*, 361, 173–194.

Osborne, P.L. (2000). *Tropical ecosystems and ecological concepts*. Cambridge: Cambridge University Press.

Osborne, C.P. and Freckleton, R.P. (2009). Ecological selection pressures for C_4 photosynthesis in the grasses. *Proceedings of the Royal Society B: Biological Sciences*, 276, 1753–1760.

Page, S.E., Rieley, J.O. and Banks, C.J. (2011). Global and regional importance of the tropical peatland carbon pool. *Global Change Biology*, 17, 798–818.

Ribeiro, K., Pacheco, F.S., Ferreira, J.W., de Sousa-Neto, E.R., Hastie, A., Krieger Filho, G.C., Alvalá, P.C., Forti, M.C. and Ometto, J.P. (2021). Tropical peatlands and their contribution to the global carbon cycle and climate change. *Global Change Biology*, 27(3), 489–505.

Sage, R.F. (2004). The evolution of C_4 photosynthesis. *New Phytologist*, 161, 341–370.

Sage, R.F. (2016). A portrait of the C_4 photosynthetic family on the 50th anniversary of its discovery: Species number, evolutionary lineages, and Hall of Fame. *Journal of Experimental Botany*, 67, 4039–4056.

Saunders, M.J., Jones, M.B. and Kansiime, F. (2007). Carbon and water cycles in tropical papyrus wetlands. *Wetlands Ecology and Management*, 15, 489–498.

Saunders, M.J., Kansiime, F. and Jones, M.B. (2014). Reviewing the carbon cycle dynamics and carbon sequestration potential of *Cyperus papyrus* L. wetlands in tropical Africa. *Wetlands Ecology and Management*, 22, 143–155.

Schlesinger, W.H. (1997). *Biogeochemistry: An analysis of global change*. San Diego, CA: Academic Press.

Singer, A., Eshel, A., Agami, M. and Beer, S. (1994). The contribution of aerenchymal CO2 to the photosynthesis of emergent and submerged culms of *Scirpus lacustris and Cyperus papyrus*. *Aquatic Botany*, 49, 107–116.

Sutcliffe, J.V. and Parks, Y.P. (1999). *The hydrology of the Nile*. IAHS Special Publication no. 5. Wallingford: International Association of Hydrological Sciences Press.

Terer, T., Muasya, A.M., Higgins, S., Gaudet, J.J. and Triest, L. (2014). Importance of seedling recruitment for regeneration and maintaining genetic diversity of *Cyperus papyrus* during drawdown in Lake Naivasha, Kenya. *Aquatic Botany*, 116, 93–102.

Terer, T., Muasya, A.M. and Triest, L. (2015). Strong isolation by distance revealed among *Cyperus papyrus* populations in the Rift Valley lakes, Lake Victoria, and isolated wetlands of Kenya. *Aquatic Botany*, 121, 57–66.

Thompson, K. (1985). Emergent plants of permanent and seasonally-flooded wetlands. In P. Denny (ed.), *The ecology and management of African wetland vegetation*. Dordrecht: Dr W. Junk Publishers.

Thompson, K. (1976). Swamp development in the head waters of the White Nile. In E.J. Rzoska (ed.), *The Nile, biology of an ancient river*. The Hague: Dr W. Junk Publishers.

Thompson, K, Shewry, P.R. and Woolhouse, H.W. (1979). Papyrus swamp development in the Upemba Basin, Zaire: Studies of population structure in *Cyperus papyrus* stands. *Botanical Journal of the Linnean Society*, 78, 299–316.

Triest, L., Sierens, T. and Terer, T. (2014). Diversity and fine-scale spatial genetic structure of *Cyperus papyrus* populations in Lake Naivasha (Kneya) using microsatellite markers. *Hydrobiologia*, 737, 131–144.

van Dam, A.A., Dardona, A., Kelderman, P. and Kansiime, F. (2007). A simulation model for nitrogen retention in a papyrus wetland near Lake Victoria, Uganda (East Africa). *Wetlands Ecology and Management*, 15, 469–480.

van Dam, A., Kipkemboi, J., Zaal, F. and Okeyo-Owuor, J.B. (2011). The ecology of livelihoods in East African papyrus wetlands (ECOLIVE). *Reviews in Environmental Science and Bio/Technology*, 10, 291–300.

2 Ancient Uses of the Papyrus Plant

Introduction

Papyrus vegetation, growing in swamps associated with the Nile was harvested and utilised for a number of uses by the ancient Egyptians from the time of their earliest settlements. In particular, the culms of the papyrus plant were used to make a range of objects that were essential for the survival of the settlers. In adapting to the regular fluctuations in water levels the swamp dwellers could construct floating homes for shelter made from a papyrus frame, covered with a thatch of papyrus culms and umbels, and also make light boats or skiffs from bundles of papyrus culms for navigating the waterways (Gaudet, 2014). Alongside these uses, ropes were made for use in the construction of boats and buildings, as well as moving heavy objects. There were also a number of everyday uses including clothing, baskets, mats, window shades and even as a food source. But it would be centuries later that the art of making the most well-known use of papyrus as a writing material was developed and eventually commercialised to the extent that it became traded internationally as a major source of income for the Egyptian state.

Unfortunately, there is surprisingly little information available about the management of wetlands used for the supply of papyrus in ancient Egypt. Some early illustrated scenes from Pharaonic times show papyrus culms being pulled from the rhizome, bundled together and tied, then carried off for use. However, the particular use to which these bundles were put is not clear. It is possible that the immature culms were used for making papyrus paper as the strips of culm in ancient papyri are of a width suggesting that they come from a less mature source. The source material for ropes on the other hand is much more likely to have been derived from sturdy mature culms which would have needed to be cut with some force from the rhizome. Furthermore, there are no records of seasons for peak harvests, perhaps suggesting continuous collection of both immature and mature culms throughout the year.

Gaudet (2014) in his book entitled *Papyrus: The Plant that Changed the World* reviewed the role of papyrus 'at the centre of one of the most vital human-plant relationships in history'. In this book he drew together

DOI: 10.4324/9781003131540-2

information from historians of ancient Egypt with his own extensive experiences of tropical Africa to highlight the role that papyrus has played in sustaining the range of ecological and economic services provided by papyrus-dominated wetlands. He makes it clear that although the production of papyrus paper and its use in recording the history of western civilisation has been the overriding interest of historians, papyrus was also of fundamental importance to the every-day activities of ancient Egyptians in providing the raw materials for making a number of essential products.

In this chapter we start with a review of the making of paper but then highlight the use of papyrus for other purposes including the construction of boats and the making of rope.

Papyrus Paper

The Paper-Making Industry

References to the plant 'papyrus' are found in lower Egypt as far back as the fourth millennium BCE and by the time of the Third Dynasty in Ancient Egypt (*c*.2700 BCE) it becomes widely illustrated in reliefs and paintings of the Theban tombs as well as in the stone columns in the stylised forms of papyrus culms and umbels (crown). Papyrus paper was used as a writing material as early as 3000 BCE in ancient Egypt, and continued to be used to some extent until around 1100 CE. Although it was probably produced almost exclusively in Egypt, papyrus (the writing material) was exported throughout the classical world. Prior to the development of papyri, written works most often appeared on pieces of wood, stone, or ostraca (shards from clay pots) (Vallejo, 2022), but papyrus became the material on which many of the Egyptian, Greek and Roman classical texts have been preserved. The earliest bible texts to have been preserved have also been written on papyrus. In the Hellenistic period (323–31 BCE), when the whole of the western civilisation wrote on papyrus, millions of papyrus rolls were made. Papyrus was also the medium of the New Testament in the early centuries after the death of Jesus. Christian texts were often in the form of a codex, rather than a roll. A codex contains several leaves bound together much like a modern book.

A review of the history of papyrus in the classical world was provided by Napthali Lewis (1974) in his book titled *Papyrus in Classical Antiquity*. Lewis intended that the book should be used 'as a tool for further research as well as a compendium of acquired knowledge', but he acknowledged how little information on papyrus (the plant) and its uses we have available to us considering how important papyrus was to the Egyptian dynasties and subsequently the Greeks and Romans. In particular, it is still remarkable how little detail we have of the earliest production of papyrus paper.

At the time that papyrus was first used for making paper it was probably found growing naturally throughout the Nile Delta wetlands and it was initially confined in its use to the Lower Nile Valley. However, in subsequent centuries, as papyrus paper use spread beyond Egypt to the Fertile Crescent (spanning modern day Iraq, Syria, Lebanon, Israel and Jordan) it may have been cultivated where climate and hydrological conditions were similar to the Nile Delta. This may have been done to ensure that security of supply was maintained when connections to Egypt were interrupted due to frequent internal and regional conflicts.

The earliest extant specimen of papyrus paper is a blank roll found in a tomb from about 3000 BCE. We know that papyrus became the principal writing material of Western civilisation, first in the Near East and eventually throughout the Mediterranean world. By the sixth century BCE papyrus was in common use in the Greek world and it was in use by the Romans in the second century BCE. Subsequently, Roman conquests carried the use of papyrus to regions in the ancient world which it had penetrated. By the fourth century BCE the Egyptian port of Alexandria exported papyrus rolls to the entire Roman world and papyrus had become an indispensable necessity of public and private business. Papyrus had clearly become the universal writing material of the Mediterranean world and it continued in use for centuries after the decline of the Roman empire.

Prior to the use of papyrus, information was recorded on clay tablets. Preserved tablets vary in size from a few centimetres to thirty five centimetres, so even if both sides were used there was not enough room for extensive texts. This was a major inconvenience when a single work was split between several pieces of clay, some of which may have been lost or broken over time. They also needed to be kept perfectly dry. To keep them safe from humidity they were stored piled up on wooden shelves or in wicker baskets. The introduction of papyrus scrolls represented extraordinary progress in the history of the book. It is suggested that the Jews, Greeks and Romans adopted papyrus paper with such enthusiasm that in the end they all considered it a feature of their own culture (Vallejo, 2022). Compared with previously available writing material the sheets of papyrus were light and flexible and on it a large quantity of text could be stored in a very small space when the sheets are rolled up. In fact the papyrus scrolls relegated tablets to a secondary status for notes and other short term uses. This innovation meant that there had been a major step forward to preserve the products of ancient thought and imagination. However, papyrus scrolls, while they survived well in the dry climate of Egypt, remaining white and flexible, they could decompose quickly in the humid climate experienced when they were exported to Europe and beyond. Nevertheless they remained the subject of a booming trade that continued to thrive until the tenth century CE. During that time the pharaohs and their successors were successful in maintaining a monopoly market by controlling access to papyrus supply and the secret of how to manufacture high quality writing material.

The eventual decline in the use of papyrus appears to have occurred quite rapidly from the tenth century, when the export trade and manufacture of papyrus from Egypt ceased. At this point, the East turned to rag paper, made by a process obtained by the Arabs from China, while the West increasingly used parchment. Throughout the long period in which papyrus was in the ascendancy the Lower Nile Valley, and especially the Delta Region, remained the principal producers of papyrus paper (Lewis, 1974). The decline in the market for papyrus meant that the papyrus swamps were no longer a source of revenue and the land that they occupied could be used for agricultural crop production. By this time the sophisticated construction of drainage channels, which had developed over the centuries, allowed the drainage of the swamps and the removal of the papyrus vegetation to be replaced by cereal crop production required to feed the expanding population.

It is generally accepted that by the time of the Napoleonic invasions in 1798, any significant papyrus swamp vegetation had disappeared from Egypt. However, in recent years its cultivated has been revived in several locations in Egypt to supply the tourist trade with freshly made examples of papyrus sheets. The quality of the papyrus sheets made from the culms is generally good as the sheets are flexible and of a creamy-white colour comparable to much of the ancient papyri. However there have been examples of dark and brittle sheets which may have been treated with caustic soda and bleaches in an attempt to artificially age the material in the hope of convincing tourists that these are ancient manuscripts (Leach and Tait, 2000). Also, it appears that manufacturers turned to another naturally occurring sedge in Egypt, *Cyperus alopecuroides*, when supplies of *Cyperus papyrus* were not available. Unfortunately this produces a poor representation of the original papyri, apparently because the vascular bundles in the culms of this sedge are less well organised than in *C. papyrus*.

With minimum variations, the papyrus scrolls were produced essentially the same way throughout their 4,000-year history. But papyrus scrolls were not only used by the Egyptians, as over time they were adopted for use by most of the societies of the ancient Mediterranean world. The wetland swamps of the Lower Nile Valley and particularly the delta region remained the principal commercial producer of papyrus writing paper and papyri, along with cereals, became the most important export product of Egypt. Over four millennia it was used to record vast amounts of evidence about the Egyptian state and in the introduction to his encyclopaedic work *Natural History* Pliny the Elder in about 70 CE wrote: 'The nature of papyrus too is to be recounted, for on its use as rolls human civilisation depends, at the most for its life, certainly for its memory.' Papyrus scrolls have provided us with the earliest extensive sources of written material in human history and have provided an understanding of the thinking and activities of the developing Egyptian, Greek, Roman and Arabic cultures. Unfortunately, few papyri from that time have survived outside of the dry areas of Upper Egypt and the Fayum region of lower Egypt, because the more generally humid climate of the Mediterranean led to the decomposition over time of the papyrus rolls. In her book *Papyrus: The Invention of Books in the Ancient World*, Vallejo (2022) tells the story of the creation of books and the history of efforts to reproduce and collect them for posterity. One of the early attempts to do this was the creation of the Royal Library of Alexandria by Ptolemy I around 300 BCE, which eventually housed many thousand scrolls of papyrus. The vigorous efforts made to continuously expand the collection included, under Ptolemy III, the requirement for all visitors to Alexandria to surrender all papyrus rolls in their possession. These were then copied by official scribes and the originals were placed in the library and the copies delivered to the previous owners. Eventually, the library became the largest in the ancient world holding at its peak up to an estimated 700,000 rolls of papyrus. However, most of the collection was lost in a fire, blamed on Julius Caesar, that destroyed the library in 48 BCE. However another factor accounting for their loss may have been the climate, as the damp and salty air of Alexandria led to the decay of the papyrus rolls even before the library building was destroyed by fire.

Information on the organisation of the papyrus paper manufacturing industry in Egypt is sparse and most of our current knowledge has been obtained from only a handful of relevant documents (Lewis, 1974; Gaudet, 2014, 2023). The discovery of the earliest scripted papyrus was made by Pierre Tallet, a French Egyptologist, when in 2013 he discovered a set of caves in limestone hills located close to the Red Sea coast. He established that the caves had served as a boat storage depot during the fourth dynasty of the Old Kingdom, about 4,600 years ago. In the caves he came on entire rolls of papyrus, some a few feet long and still relatively intact, written in hieroglyphics as well as hieratic, the cursive script the ancient Egyptians used for everyday communication. The papyri were written by men who participated in the building of the Great Pyramid, the tomb of the Pharaoh Khufu, the first and largest of the three colossal pyramids at Giza just outside modern Cairo. Among the papyri was the journal of a previously unknown official named Merer, who led a crew of some 200 men who travelled from one end of Egypt to the other picking up and delivering various goods. The first clear indication that the use of papyrus had spread beyond the borders of Egypt is found in a text of circa 1100 BCE. This referred to the 'Voyage of Wen-Amon', who travelled from Egypt to Phoenicia, taking with him 500 rolls of very fine papyrus. By the sixth century BC, papyrus was in common use in the Greek world. Direct relations between Rome and Egypt existed from the reign of Ptolemy II (died 246 BCE) and it is assumed that Roman conquests subsequently carried the use of papyrus to the regions of the ancient world to which it had not previously penetrated. Papyrus rolls, exported from Egypt, had become a necessity of public and private business to the entire Roman world. Once established as the universal writing material of the Mediterranean world, papyrus continued in use for centuries after the decline of the Roman empire. The conquest of Egypt by the Arabs in 641 CE did not immediately affect the papyrus trade. However, interruption in supply usually resulted as a result of the intermittent revolts of the Egyptians against their new rulers.

Some of the largest and most productive papyrus swamps to supply the papyrus paper manufacturing industry were in the Fayum Oasis, a region of ancient Egypt known for its fertility and the abundance of plant and animal life. The Fayum Oasis is located about 100 km southwest of Cairo in a low depression to the west of the Nile and surrounded on all sides by the Libyan Desert. It was once an arid desert oasis but it was formed when a branch of the Nile River silted up and diverted water to it in around 7200 BCE (Williams, 2019). The region was at its most prosperous during the middle kingdom (2030–1640 BCE). It was the preferred location for hunting wild animals by the Egyptian nobility and papyrus grew in abundance. In the early Middle Kingdom, Amenemhat I ordered the construction of a canal fed by the Nile which flooded the north west region of the Fayum Oasis and created the great Lake Moeris with an estimated area of about 1,500 km^2. The Fayum engineering project gave the pharaohs the power to begin regulate the Nile, prevent destructive floods and provide precious water relief in times of drought. It turned the barren desert into Egypt's granary. The seasonal flooding supplied irrigation to the surrounding area making it one of the most fertile parts of Egypt. It was in the permanently flooded backwaters, where the papyrus swamps developed, that the papyrus plants were harvested.

The wetlands of the Nile Delta and the Fayum region appear to have supplied virtually the whole ancient world from Britain to Mesopotamia with the only writing material which was at once convenient, durable, and reasonably priced. Lewis (1974) in *Papyrus in Classical Antiquity* concluded that 'throughout classical antiquity the swampland of the Lower Nile Valley and especially of the Delta region, remained the principle, and in all probability the only, commercial producers of papyrus and its products'. However, information on the organisation of the papyrus industry is sparce. Lewis (1974) refers to a handful of relevant Greek documents complemented by isolated bits of supporting evidence. He states: 'the sum total presents us with a chiaroscuro, a few highlights emerging from a dark background of silence'. What we do know is that throughout several centuries of classical and late antiquity, Egypt supplied the everyday needs for writing paper. As we have seen, the papyrus plant's natural distribution in the Lower Nile was probably confined to continuously flooded backwaters. These were primarily a number of depressions of variable size distributed across the Lower Nile Valley including the Fayum Oasis and the Nile Delta. It is likely that large parts of these areas were reclaimed over time for agriculture by a system of irrigation channels, but the areas of 'natural' wetland remained for harvesting of papyrus. For villagers these areas were convenient places to pasture and water cattle around the perimeter while the central permanent swamp was both a source of fish and game, but primarily the 'quality' papyrus culms that were required for making papyrus paper.

For making the higher quality papyrus paper, the papyrus culms were not allowed to grow to full maturity as can be seen from the width of the strips that have been used in the making of preserved papyrus paper. This is because older culm tissue discolours, turning brown a short time after harvesting; probably due the increase in tannin content. It is therefore safe to assume that 'quality' paper required much 'softer' and less lignified tissues (Bausch et al., 2022). The harvesting of papyrus is depicted on the walls of Egyptian tombs and they illustrate culms being pulled from their bedding in the rhizome. This supports the view that culms were harvested when they were still immature, because it becomes more difficult to extract culms in this fashion as they mature, when there would be a need to resort to cutting at the base to harvest them. Judging from recent experiments where the growth of papyrus culms has been carefully monitored, they probably become fully mature in 8–9 months (Jones and Muthuri, 1997). It is therefore likely that the papyrus culms for paper manufacture were harvested about six months after they first emerged.

Because papyrus produced for export was a very valuable source of revenue for the government it is likely that suitable areas of swamps were carefully managed to produce high yields and culms of the quality to make marketable papyrus paper. Because the culms, soon after harvesting, rapidly begin to dry and discolour, there is need to process them quickly. Consequently, the factories where papyri were made must have been very close to the swamps where the papyrus was growing. This would have ensured a daily delivery of freshly harvested culms. Considering the large number of papyrus rolls produced, these factories would have employed considerable numbers of workers. Surprisingly, once again, we appear to have very little direct evidence of the size and location of these factories. It is probable that papyrus was manufactured throughout the Nile Delta and Fayum region as factories of varying size must have been set up wherever the raw material was available. Furthermore, in order to produce the highest quality of raw material, the cultivation of the papyrus must have been carefully managed to ensure a stable supply for both a large domestic and export market. The paper manufacturing probably took place year-round, although June, July and August appeared to be the most productive time of the year when the papyrus swamps were most accessible due to the lowest water levels. From September to March the floodwater of the Nile would be at its highest and the access to the papyrus plants most difficult, probably requiring the use of boats. The bundles of papyrus culms were delivered and marketed in units of one armful and transported to the site of paper manufacture in armful loads. There is evidence that under the Ptolemies (305–30 BCE), the last dynasty of ancient Egypt, before it was incorporated into the Roman Republic when production of papyrus rolls for export were reaching their peak, their marketing appears to have some level of government control through supervision by government agents. Lewis (1974) suggests that, like production of other essential goods like linen, that factories were required to deliver quantities of acceptable quality to the government authorities, after which they were allowed to produce for independent sale. However, this would only be under appropriate conditions that did not impact the state supply. Under subsequent Roman rule there appeared to be a less controlled economy which meant that factories were largely under private ownership, although there may have some limited degree of state control and supervision.

A question frequently asked is 'was papyrus a cheap commodity, or a dear one?' Once again there is great uncertainty, with scholars finding evidence for both. The expensive side of the argument is supported by the frequent use of the back of the papyrus sheet, often for important records, and the continued common use in Egypt of ostraca, which cost very little, that were used widely as 'notepads' for private letters and copies of literary works. Furthermore, the existence of the Ptolemaic papyrus 'monopoly' probably kept prices high. The cheap side is supported by the apparently plentiful supply and wide use of papyrus, even as wrapping paper. Although this paper is usually of a poor grade, unsuitable for writing on, or made up of discarded documents. Clearly, an early example of recycling! Also, large amounts of poor quality and previously used papyri were used for making mummy casings, which consisted of layers of papyrus built up and bound with an adhesive such as glue or starch. Unfortunately, hard facts and statistics are missing but it has been suggested that there was a decreasing price as a result of increasing centralisation of production. Lewis (1974) suggests that during most of classical antiquity a roll of papyrus cost the equivalent of one or two days' wages, although it was possibly as high as what a labourer might earn in five or six days. Clearly, the unskilled labourer would have had little need of writing paper, which were beyond his means. However, for the higher social classes it is unlikely that these costs would have been prohibitive.

Although papyrus paper had been manufactured and used widely in ancient Egypt it was not until the Greek and Roman times that its use probably peaked. Statistics on total production and consumption do not exist, so we are left to draw assumptions on a relatively small amount of information that has been gleaned, largely by Lewis (1974), from a small number of random sources. For example, he refers to information that in the middle of the third century BCE, a large estate in Egypt purchased papyrus rolls in lots of a thousand or more that were distributed to their accounting offices. These offices were using as many as twenty rolls per day. This was indicative of prolific domestic consumption while at the same time there was a developing export market. Two contracts for private leases of papyrus plantations from the first century BCE indicate that while natural, 'unmanaged' papyrus wetlands yielded materials for matting, sandals, matts and the like, there were specially managed plantations that met the massive demand for quality writing materials across the Roman Empire. Management probably involved the regulation of harvest intervals of the papyrus culms and controlled access routes to the swamp. However, there were seasonal differences in supply of papyrus culms to the factories which may have reflected differences in availability of papyrus to harvest. In one contract the lessees paid double rent for the period March to August, implying that production peaked in the summer months.

It is estimated by Lewis (1974) that, at the peak demand for papyri, millions of rolls were produced each year, virtually all of this in Egypt and exported via Alexandria, which became the head-quarters of the papyrus market. As Mediterranean societies became literate and more complex they needed more and more papyrus, and prices rose accordingly to meet the feverish demand. Outside Egypt the cultivation of papyrus and manufacture of papyrus paper was very limited and, like coltan used in current day smartphones, papyrus became a strategic product. A powerful market therefore developed, distributing papyrus on commercial routes through Asia and Europe.

It is generally accepted that from the tenth century CE onwards there was a catastrophic decline in the export market for papyrus paper. The decline in the market was probably due to a combination of competition from other writing materials as well as an inability to ensure a reliable supply of papyrus scrolls as the demand increased. This time corresponded with the further decline of the Egyptian dynasties and the cause appears to have been a combination of a century's long drought, a series of economic crises and opportunistic foreign invasions. The initial competition from other writing materials came from parchment and then rag paper. Parchment is made from specially prepared untanned skins of primarily sheep, calves, and goats. Its advantage as a writing surface was that it was smoother than papyrus paper and more durable and resistant to mishandling. It was also more resistant to the damaging effects of humid environments, especially in moister northern climates. The replacement of papyrus by parchment probably also reflects the fact that parchment can be manufactured anywhere, as its raw material is not as geographically restricted as the papyrus plant. The invention of paper (c.105 CE) is traditionally ascribed to Cai Lun, a eunuch at the court of the Han emperor Wu Di. Paper was originally made from textile rags from hemp, linen and cotton and by the eighth century CE it had replaced papyrus, and was more widely used than parchment, as the preferred writing material as far west as Sicily and Spain. The invention of printing in the 1450s brought a massive increase in demand for paper which could never have been met by papyrus production based largely in Egypt.

As the progression to parchment first and then paper progressed the demand for papyri fell and consequently the controlled management of the papyrus swamps in Egypt also declined Eventually, papyrus vegetation disappeared from the Lower Nile region and it was not until a retired Egyptian engineer and diplomat named Hassan Regab brought cuttings of papyrus back from Sudan in the 1970s that papyrus was established again at an Institute in Cairo where Regab subsequently made papyrus paper for the tourist trade. Papyrus swamps appear to have had a very limited distribution in the Nile Valley following the decline in its use for making papyri so much so that by the time of the French expedition of 1798–1801 CE under

Napoleon Bonaparte it appears to have been extinct. Any remaining swamps growing papyrus had little commercial value so that they were drained to meet the demand for land for growing food crops.

Papyrus Paper Manufacture

Writings on the relative merits of papyrus, parchment and paper have frequently dismissed papyrus as a rougher, darker and more brittle than the other two material. However, papyrus produced by the ancient factories had, and retained for many years a surface that was shiny and smooth and was flexible. Although papyrus does, like pulp paper, yellow with age and become more brittle, papyrus rolls were repeatedly rolled and unrolled for decades and even centuries. In fact, in ancient and medieval times papyrus rolls and documents may have had a usable life of hundreds of years (Bausch et al., 2022).

Unfortunately, no ancient source of information has provided an unequivocal description of papyrus production that has allowed current researchers to straightforwardly reproduce the process. As a result the art of papyrus making remained largely unknown for more than a thousand years. During the twentieth century, when more and more papyrus texts came to light, scholars began to investigate how ancient papyrus manufacturing may have occurred. The most extensive early historical account of papyrus production had been found in the encyclopaedic work by Pliny the Elder (23–79 CE) called *Naturalis Historia*, written in about 77 CE and most research has subsequently used this account as a starting point (Lewis, 1974; Leach and Tait, 2000; Flieder and Delange, 2001). Although there have been many discussions about the veracity of this description the general consensus now is that it is largely correct in basic methods but ambiguous in detail.

Basically, papyrus sheets are made by arranging two layers of sliced papyrus culm pith, one on top of the other, at right angles. The layers are then pressed together to bond the strips of pith and form a sheet of papyrus. Initially the papyrus culm can be cut into long continuous strips made up of vascular bundles and simple parenchyma cells that give body to this framework. The walls of the vascular bundle cells contain fibres of cellulose, made up of long chains of glucose molecules, which provide the structural strength for the body of the papyrus sheets (Mohamed, 2023).

The sides of a papyrus sheet where the vascular bundles run horizontally are the recto and, where the fibres run vertically, the verso. The recto was preferred for writing on but the verso was used for documents as well, allowing two separate texts to be included on a single papyrus. According to Pliny, no glue was added for the adhesion of the strips, but sheets made with the lower part of the culm are more translucent and stick together better than sheets from the upper part of the plant. The lower part of the culm is thicker and as a result the vascular bundles are more widely spaced than the upper part of the culm. This is because the vascular bundles run the length of the culm, which tapers towards the top meaning that the vascular bundles are more tightly packed at the top, making this part of the culm more fibrous. While the broad scheme for manufacture of papyri is clear, the *exact* method of papyrus-making was undocumented by the ancient Egyptians, and therefore some of the details of the procedure have been pondered over by modern scholars. Several variations on the basic scheme outlined by Pliny the Elder have been proposed and tested, but none has produced a writing material which is of the highest quality of ancient papyrus (Leach and Tait, 2000).

Papyrus paper production started with the selective harvesting of suitable culms in the swamps, presumably close to where they would be processed. After the culms were gathered and while they were still fresh and green, a 20–48 cm section of the culm was cut (the maximum length being the maximum height of a papyrus scroll) and the outer rind of the culm, made up of the epidermal and sub-epidermal cells, was stripped off. The tough green rind of the papyrus culm encloses the white pith, from which the papyrus sheet is made. The pith is composed of the sparsely arranged rigid ligneous and cellulosic vascular bundles, and parenchyma cells with cellulose and hemi-cellulose walls, with spaces between them that function as air passages. The morphology of the papyrus pith is what lends a papyrus sheet its characteristic criss-cross pattern. The vascular bundles are the more substantial striations running horizontally across the recto of a papyrus (and vertically on the verso), and the parenchyma cells are the paler, more delicate 'filling' between the striations. Thin strips were about 2–4 mm thick and produced by slicing the pith lengthwise. Clearly, great skill was required to create thin and uniform strips. Some researchers have suggested that an alternative interpretation of Pliny's description of the processing of the culms was done by peeling the soft pith continuously down the culm in an unrolling action that results in one continuous sheet. However, modern attempts to reproduce the process have shown that it is even more difficult to peel the pith using this method so that the layers are uneven when pressed together (Wallert, 1989). Individual sheets were formed by laying the strips side by side, first in one vertical layer, and then in a second horizontal layer. There are some uncertainties about whether the strips were butted or overlapped and observations on ancient material suggest the either was used. However butting up avoided ridges where the two layers overlapped so this was probably favoured in the highest quality papyri. It is generally accepted that no adhesive was used, and although the papyrus cell sap contains simple sugars including galactose and arabinose which are the main constituents of natural adhesives such gum Arabic, it is the pressing together of the strips that is required to ensure physical bonding of the layers (Leach and Tait, 2000). In recent attempts to reproduce the process, it has been found that using pre-soaked strips of papyrus leads to better adhesion between the layers by facilitating physical bonding, as the moisture causes the parenchyma cells to swell and then lock into each other upon drying (Bausch et al., 2022).

Using electron microscopy it has been found that the microfibrils which make up the cell walls are arranged in a series of spirals, which enables a highly effective bond to be made as the microfibrils interlock together under pressure. Modern researchers have found that pressing until dry is critical for sheet formation although Pliny claims that sheets were pressed together and then allowed to complete drying in the sun. The exact method of pressing and drying is unknown and as there is no evidence from Pharaonic Egypt for any kind of press the possibility still remains that the skilful beating with mallets was the preferred method to ensure effective binding of the papyrus layers. A consequence of the process of pressing expels large amounts of air in the intercellular spaces and as a result the paper is reduced in thickness to less than a quarter of the original thickness (Leach and Tait, 2000). The colour of the papyri when dry appears to be related to the thickness of the strips and how fresh the culms are. The thinner the papyrus the quicker it dries and the whiter it is. Mohamed (2023) has suggested that papyrus sheets were treated with various different materials, including egg white and dilute starch, to improve the surface for writing by preventing the over-absorption and spreading of the ink. He also suggests that cedar oil extracted form cedar trees protected the papyrus from microbial and insect damage. The oxidation of the plant polyphenols in the presence of oxygen to produce quinones, which are coloured, is the reason why papyrus turns brown in contact with the air. However, drying deactivates the enzymes involved and halts the process of discolouring. Bausch et al. (2022) have investigated further the yellowing of papyrus sheets that occurs over time. They compared the chemical composition of ancient papyri and papyrus sheets that they had produced themselves following as closely as possible Pliny's protocol. Their particular interest was in the role of polyphenol oxidation in determining the discolouring of papyri. Analysis showed that papyrus culm pith is composed of approximately 80% cellulose/hemicellulose and 20% lignin, which is a complex organic polymer consisting of cross linked polyphenols. They concluded that the ancient papyri were white when first manufactured and over time became yellow as the polyphenols in the lignin slowly oxidised, particularly under humid conditions.

Sheets of papyrus were traditionally made into rolls; a common size is twenty sheets to a roll, although occasionally individual sheets were sold for record keeping and lists. Rolls were made by overlapping sheet edges by 1–2 centimetres and adhering with a flexible starch-based paste. Pliny states that the paste used was made from the finest flour and allowed to stand for a day. When the joints had been pasted down they were flattened and smoothed with a mallet to beat the overlapping joints of four layers of papyrus strip to a thickness similar to the rest of the roll. Once the papyrus sheets had formed a roll they were rubbed smooth with ivory or a shell (Lewis, 1974). Mainly blank rolls were produced, but occasionally already inscribed papyri sheets were added to previous rolls, or adhered together to make a new roll, among other alterations. Papyri were always rolled with the vertical-running pith on the outside because of the natural tendency of the sheet to curve in that direction; inscriptions and illustrations on a papyrus roll are generally on the horizontal side of the sheet, where they would be protected when rolled up. The rolls were then transported to temples, government buildings, the market or exported in trade. John Gaudet (2014) in his book on papyrus, makes an estimate of the production of papyrus paper in Ancient Egypt, when papyrus paper use was at its peak during the Roman times, to be about 50 million sheets per year. In practice paper makers leased swamps and their management was regulated to ensure sustainable production. Papyrus was of course also needed for other purposes but the quality of the papyrus was not as crucial as it would have been for making paper. For the bulk of the Egyptian population there was little use for papyrus paper but there were other multiple uses which were important to their everyday life. Also, the transition to an afterlife was normally undertaken wrapped in a papyrus mat and clutching a tiny amulet containing a small piece of papyrus paper.

As a writing material, papyrus was used for hymns, religious texts, spiritual admonitions, letters, official documents, proclamations, love poems, medical texts, scientific or technical manuals, record-keeping, magical treatises, and literature. Extant scrolls range from fragments of one page to the famous Ebers Papyrus dated to about 1550 BCE, which is 110 pages on a scroll 20 metres long. The Ebers Papyrus is a medical text which is routinely cited as evidence of how medicine and magic were interrelated in ancient Egypt. Along with other papyrus scrolls like the Kahun Gynaecological Papyrus, the oldest medical text in Egyptian history, the London Medical Papyrus, and the Edwin Smith Papyrus, these works attest to the vast medical knowledge and skill of the ancient Egyptians and how they went about addressing major and minor injuries, various ailments, and serious conditions such as cancer and heart disease. Cases of anxiety, depression and trauma are also dealt with in the medical texts of Egypt, as are subjects like abortion, birth control, menstrual cramps and infertility.

Ancient Egyptian scribes wrote in black and red ink. Red was used for the names of demons or evil spirits, to mark the beginning of a new paragraph, for emphasis of a word or passage, and for punctuation in some cases. The source of the red ink was haematite, an iron oxide mineral. The black ink was made from carbon obtained by partially burning organic material such as wood and mixing it with gum such as gum Arabic. Scribes carried a wooden case which held cakes of black and red paint and a water flask to mix and dilute the paint into ink. The pen was initially a thin reed with a soft tip but was replaced in the third century BCE by the stylus, a more robust reed sharpened to a very fine point. A scribe would begin a work on the recto of the papyrus scroll, write until it was filled, and then flip it over to continue the text on the verso. In some instances, a papyrus roll on which only the recto had been used would be taken by another scribe and used for another work, either complementary or completely unrelated. The image of the Egyptian scribe hunched over his papyrus scroll is accurate, but long before he got his hands on that scroll, he would have spent literally years practising writing on potsherds, chunks of stone, and pieces of wood. The scribes of ancient Egypt spent years learning their craft and, even if they were from wealthy families, they still were not allowed to waste precious material on their lessons. Only when they had mastered the basics of writing were they allowed to practice on a papyrus scroll. Almost nothing is known about the way in which a Pharaonic scribe obtained his writing material. He may have been allotted papyrus by the sheet or have bought it from the state or from a papyrus manufacturing workshop. At all periods the re-use of papyrus was widespread, which suggests that it cannot have been cheap. Brukner Havelková et al. (2024) examined skeletal remains of scribes to determine whether repetitive tasks and maintained postures related to scribal activity manifested itself in skeletal changes and identify possible occupational risk factors. Changes in the skeletons of scribes showed an enhanced occurrence of osteoarthritis of the joints. Their research showed that remaining in a cross-legged sitting or kneeling position for extended periods, and the repetitive tasks related to writing caused extreme overloading of the jaw, neck and shoulder regions.

Survival of Ancient Papyri

Although papyrus paper, like any other organic material will decompose over time, largely due to microbial and insect attack, it is remarkable how well preserved are some papyri found from as far back as 3000 BCE. However, these are very rare examples and the vast bulk of the millions of sheets of papyrus that were produced and inscribed over the centuries have long ago decomposed and their carbon content has been recycled to the atmosphere as carbon dioxide and methane. In fact, the most well-preserved material has been found in tombs within the confines of the modern country of Egypt where their durability has most likely been aided by the favourably dry climate. Until discovery these tombs have retained a low and stable relative humidity and stable temperatures but probably most significantly they have been protected from human interference. Less well preserved papyri have also been excavated from archaeological sites in a wide variety of fragmentary states due to decomposition caused by a damp atmosphere and fluctuating temperatures. However, preservation at settlement sites have been better when atmospheric conditions have been relatively dry due to a low water table (Leach and Tait, 2000).

Multipurpose Papyrus

Although the majority of the population of the Egyptian Kingdoms would not have been literate and had no use for papyrus as a writing material, the harvested culms were put to a number of other uses in their everyday life including making basketry, sandals, mats and basic furniture (Leach and Tait, 2000). It is likely that Egyptians found uses for every part of the papyrus plant, from the roots and rhizomes to the crowning umbels. Mats and window shades were woven through a technique similar to that used to make writing material but in this case the part of the culm used was the external ring of 'rind' around the perimeter. The rind was stripped and laid down vertically and then woven with others horizontally and pulled tightly; they were then bound with a thinner strip of fibre from the plant. Sandals were made by coiling the papyrus and were so sturdy that many examples of them have been found thousands of years after they were made, still in good condition. Papyrus sandals required a great deal of skill to make and were too expensive for most people. The Greek writer Herodotus, writing in the fifth century BCE, reported that the priests of Amun, the highest ranking priests in the priesthood of the ancient Egyptian god of Amun, only wore papyrus sandals which, along with other evidence, scholars interpret as proof of the priests' great wealth. Dolls or other toy figures were also made from young papyrus by bunching up the developing culms and then shaping them using tightly tied fibres to create a head, arms, and legs.

The variety of uses of papyrus by ancient Egyptians testify to its utilisation in modern day terms as a 'sustainable multipurpose crop'. The culms could be harvested at regular intervals as after cutting a new population develops to replace those that have been removed. In addition to the production of papyri and the every-day local uses of papyrus, two other extremely important uses were at the centre of the development of the ancient Egyptian society. These were the construction of early river craft and the production of ropes.

Papyrus Boats

The development of Egyptian civilisation would not have been possible without riverine navigation (Hassan, 1997). During a long period of prehistoric time, until wooden boats arrived, boats made from papyrus were essential for people to make use of the river. The earliest river travel was on simple rafts or skiffs made of bundles of papyrus culms and these were probably used as early as 7000 BCE to fish the main channels of the Nile (Ward, 2006). These simple boats would also have been used to transport people between settlements during the annual floods and also for the transfer of goods, particularly food crops, from one place to another. Boats therefore played an important role in unifying the country as food could be transported from one district to another to relieve famine and support people in the whole region to overcome shortages in any district. Besides the donkey, which was used for overland transport, boats made possible the economic integration of the country (Hassan, 1997). It has been suggested that the emergence of kingship in Egypt and the development of administrative units, may have come about as a consequence of the coordination of the collection of grain and relief of shortages to deal with unpredictable crop failures in any of the districts. As a result towns eventually emerged in the administrative units as cult centres and capitals of political power (Hassan, 1997).

The earliest Egyptian light weight skiffs were made by tightly weaving stalks of papyrus and binding them with rope, also made of papyrus. To make the papyrus buoyant bundles of culms were tied together in several places to trap the air in spaces in the culms. This technique created a lightweight water-tight boat which could easily be carried by hunters or fishermen until they needed to be launched. They were paddled, like a canoe, or propelled with a long pole, like a punt, and they offered an easy and inexpensive solution for a short journey. The papyrus skiff is featured in numerous tomb and temple paintings and has a markedly different, more linear, shape than later wooden boats built on the same design.

Papyrus boats played an important role in the myths, legends, religious ceremonies and pageants of the pharaonic Egyptians. Their passage into the next world was by a boat made from papyrus culms and this frequently featured in Egyptian art. Tomb scenes also show high-ranking members of ancient Egyptian society using their skiffs for leisure pursuits such as fishing or fowling in the wetland marshes along the banks of the river. Illustrations show the boats of two principal shapes, one with the prow and stern raised up high and the other with only slightly raised extremities. The latter probably accommodates the use of square sails, which were also made of papyrus fibres. These boats were light weight and portable and were ideal to service the extensive marshes of the Nile Valley. Ancient Egyptian vases and murals from around 6000 BCE depicted larger boats that could even take to the sea, built with papyrus culms but propelled by rowers using oars. In 1969 a Norwegian, Thor Heyerdahl, suggested that papyrus boats could have navigated the oceans and in order to demonstrate this he proposed building a boat large enough to sail across the Atlantic. Heyerdahl was an explorer and scientist who had risen to international fame following his 1947 *Kon-Tiki* expedition and the subsequent bestselling book and Oscar-winning film of the same name. The purpose of the *Kon-Tiki* expedition was to prove that migration from South-America to Polynesia could have taken place in pre-colonial times by crossing the Pacific Ocean on a raft. Heyerdahl later took the initiative in 1969–1970 to lead another expedition of similar scope with papyrus boats named after the Egyptian sun god, they were named *Ra I* and *Ra II*. He did this because he had become interested in the possibility of cultural contacts between early peoples of Africa and Central and South America. He suggested that certain cultural similarities, such as the shared importance of pyramid building in ancient Egyptian and Mexican civilisations, perhaps indicated a link. To test the feasibility of ancient transatlantic travel, Heyerdahl built a copy of an ancient papyrus vessel in 1969, with the aid of traditional boat builders from Lake Chad in Central Africa. Because papyrus no longer grew in Egypt, the boat was constructed from tons of papyrus shipped from Lake Tana in Ethiopia, and was eventually transported to Safi in Morocco, from where it set sail for the Caribbean in May 1969. However, defects in the design and other problems caused it to founder in July, after 3,000 miles, 600 miles short of its goal. Undaunted, Heyerdahl constructed a second papyrus boat, the *Ra II*, with the aid of Aymaro Indian boatbuilders from Lake Titicaca in Bolivia. With a multinational crew of seven, the *Ra II* set sail from Safi on May 17, 1970. After a voyage of 57 days and 4,000 miles, the ship arrived in Barbados.

During pharaonic times, by about 3250 BCE, bigger and more practical river boats were being constructed as trade along the Nile was flourishing and there was a demand for stronger, roomier cargo vessels. As a result, Egyptian boatbuilding changed from construction with papyrus culm bundles to wooden planked vessels. However, as forest supplies were limited and costly, for everyday use papyrus boats continued to be built. The first wooden boats were modelled after papyrus boats with flat bottoms, no keels and square sails. The transition was documented in drawings on pottery of large vessels travelling along seasonal watercourses (Ward, 2006). The long, curved, boat structures typically featured paddles, steering oars, upright and curved ends. The wood used was mainly *Acacia*, which was imported from the Near East as part of an intense exchange of trade. The boats were built in carvel fashion by laying planks edge to edge to form a smooth outer surface and stitched together with papyrus rope. The spaces between the planks were caulked with papyrus, which involved crushing strips of fresh papyrus culm stems between the planks of wood. This form of construction allowed the boats to be disassembled and the planks transported, reassembled and recycled. Ward (2006) suggests that this allowed the dismantled boats to be carried over land to the Red Sea where they could be reassembled in order to sail along the Red Sea coast to trade for aromatic resins, gold and other commodities important to the socially competitive Egyptian hierarchy. The use of papyrus in the construction of these boats ensured that it continued to be a significant component of Egyptian boatbuilding even after wood replaced it as the primary material for construction. Furthermore, when small wooden vessels were developed into large sailing ships, the papyrus woven into ropes was used extensively on the boats and in the rigging. The first sails were also made from the woven stripped epidermis of papyrus culms but these were superseded by linen cloth woven from the bast fibre of flax.

Papyrus Rope

The ancient Egyptians were probably the first civilisation to develop special tools to make rope (Teeter, 1987). The use of such ropes, often pulled by thousands of workers, allowed Egyptians to move the heavy stones required for the building of their monuments as well as other heavy duty tasks. Papyrus ropes have been described as the 'workhorse of ancient Egypt' as they were able to support the heaviest of undertakings. During most periods of pharaonic Egypt, it was customary for non-royal tombs to be decorated with scenes of daily life and industry and these have been invaluable in understanding and reconstructing the techniques employed by ancient craftsman to make ropes and how the ropes were used (Teeter, 1987). The

'Rope Cave' uncovered at Gawasis on the Red Sea 23 km from the modern port of Safaga, contained a large number of well-preserved rope coils. The ropes were most likely used on ships of the state-sponsored maritime voyages of the middle kingdom from about 2030 BCE to 1640 BCE (Borojevic and Mountain, 2011). As papyrus does not grow near the sea in saline soils it is likely that the ropes described here were manufactured and brought from the Nile Valley, more than 150 km away. Almost identical ropes have been found in other locations in ancient Egypt and it is likely that papyrus was the plant of choice for making large, thick ropes in particular. Gaudet (2014) suggests that papyrus rope was both an export earner and was made to supply the large market for papyrus rope in the Egyptian ship building trade, so that enormous quantities were produced every year. For example, Gaudet (2014) estimated that the typical wooden ship of the Nile required about 300 m of papyrus rope in her rigging and even more rope was required for similar size papyrus boats.

Borojevic and Mountain (2011) carried out a very thorough anatomical study of the plant material used for the manufacture of the ropes found in the 'Rope Cave'. They showed that the ropes are made from three strands, each with multiple interior yarns, indicating it was constructed from plant stems or culms rather than leaves. The most prominent features examined were epidermal ridges, as well as the internal tissues and vascular bundles visible in cross section. The epidermal ridge patterns most clearly resembled those of *Cyperus papyrus*. In longitudinal section the cellular structure was also closest to papyrus but different from other *Cyperus* species such as *C. rotundus* and *C. alopecuroides*. In addition, the ropes showed no evidence of nodes, which are characteristic of reeds such as *Phragmites australis* and the giant reed *Arundo donax*. This research clearly cast doubt on earlier, less detailed, analyses of the rope material which had suggested that the Cave ropes had been made from *Phragmites communis*. A particularly important reason to support the view that papyrus was the plant of choice for making large thick ropes was that it is a more pliable material than the stiff stems of *Phragmites* or *Arundo* (Borojevic and Mountain, 2011). In addition, more evidence supporting the extensive use of papyrus in rope making comes from a number of tomb paintings dating from as early as the Fifth Dynasty (*c*.2465–2353 BCE) and as late as the Eighteenth Dynasty (*c*.1550–1295 BCE) showing rope making taking place in papyrus swamps (Teeter, 1987).

Manufacture of papyrus rope probably took place in areas of swamp separate from those used for the cultivation of papyrus for making paper. The papyrus for paper probably required relatively young culms which were carefully harvested, while the manufacture of large ropes required culms from mature papyrus from which long strips of skin or rind could be peeled. The ropes were made employing one basic technique which involved twisting and counter twisting the strips of fibre from the papyrus culm. The process involved first drying the harvested culms and then soaking them in water to make them more pliable. The peelings could be further sub-divided to make long strips of a strong raw material from which finer rope or string could be made. However, the thicker ropes used in moving large loads and on ships probably consisted of multiple strands of fibres from whole papyrus culms. The ropes were made by making strands of fibres first twisted clockwise to the right and then combining at least three strands by twisting them in the opposite direction to produce a finished rope. The making of these ropes was highly organised on an industrial scale. Teeter (1987) has suggested that the use of heavy swinging weights attached to the fibre strips to make long lengths of rope was most probably needed to twist the tough fibre strips which could not be twisted by hand.

Papyrus as Animal Forage and a Human Food Source

The Egyptians herded cattle and sheep and kept goats from Predynastic times. The floodplains were extensively used for the cultivation of cereals and the herding of grazing animals was most likely confined to the Delta marshlands and along the inner wetland margins of the floodplain (Hassan, 1997). The crude protein content and ruminant dry matter digestibility of the umbels of papyrus have been shown to be similar to those for a range of grasses growing in these locations but the large tough culms are certainly less palatable (Muthuri and Kinyamario, 1989). It is possible that during the dry seasons, when the wetland margins dried out as water levels fell and the grasses stopped growing, an alternative source of feed could have been papyrus umbels flourishing further into the wetland. There may have been times therefore when papyrus provided a source of animal forage after harvesting the umbels but it is likely to have been only during periods of severe drought when the annual floods were reduced or failed all together.

There is also some evidence that papyrus was a source of food for the early settlers along the Nile. Herodotus suggested that the rhizome of the plant could be roasted and eaten and was a staple of the Egyptian diet. However, this appears to be unlikely for societies in later dynasties as analysis of stable isotopes of carbon in hard and soft tissues from Egyptian mummies, used to indicate the major source of food to construct the tissues, indicate that the major part of the diet was from plants with C_3 photosynthesis rather than the C_4 photosynthesis that would indicate the food source was papyrus (Touzeau et al., 2014). This suggests that the bulk of the diet was much more likely to be cereals such as emmer wheat and barley produced by plants with C_3 photosynthesis. Because barley is more resistant to salinity and fluctuations in groundwater, it was probably cultivated more widely than emmer wheat (Hassan, 1997). It is possible that the very earliest settlers along the Nile were more dependent on the extensive papyrus swamps as a source of food but this would have changed as cereal cultivation developed to form the staple food source of the entire Egyptian civilisation. However, papyrus may have provided some minor nourishment as it was observed by Theophrastus that 'all the people in the countryside chew the papyrus stalk, raw, boiled or baked; they swallow the juice and spit out the pulp' (Lewis, 1974). It was the annual flooding of the Nile and the presence of fertile soils enriched by the deposition of silts carried in the flood water that allowed Egypt to eventually become the bread basket of the region. It was likely that the developing agriculture, based largely on cereal production, was a major reason for the clearing of many of the papyrus swamps and those that remained were used as a source of materials primarily for the production on an industrial scale of papyri, boats and ropes. However, for rural populations that lived close to the papyrus swamps they must have continued to harvest papyrus culms for use in their everyday life, including making basketry, sandals, mats and basic furniture.

Conclusions

From the time of the earliest settlements of the Nile Valley, papyrus had been harvested for use by the local population. Initially the marsh dwellers probably harvested papyrus to support the provision of basic requirement for survival in the swamp regions. Shelters could be made from papyrus culms and it is likely that these were built on the floating papyrus mat, so that they rise up during the annual inundation of Nile floodwaters. These early marsh dwellers also used papyrus to make basic furniture as well as mats, containers and primitive skiffs for moving around the swamp which was almost impossible to penetrate by foot. It was centuries later that the art of making the most well-known use of papyrus, as a writing material, was developed and eventually commercialised to the extent that it became traded internationally as a major source of income for the Egyptian state. The preservation of many of these papyrus scrolls have enabled historians to elucidate some of the earliest developments of western civilisations as far back as the fourth millennium BCE. The history of papyri is remarkable as the scrolls were produced almost exclusively in Egypt and eventually contributed to a massive export industry as the demand for this writing material expanded throughout the middle east and Europe. Despite the significance of papyri for Egyptian society and papyrus swamps being the setting for some of the most important of the ancient religious legends of the Egyptians there is little recorded information on how papyrus swamps were managed and how they were sustainably maintained, not only to produce papyri but to provide material for other essential societal needs including boats and ropes. It is still remarkable how little detail we have of the earliest production of papyrus paper. Basically, papyrus sheets are made by arranging two layers of sliced papyrus culm pith, one on top of the other, at right angles. The layers are then pressed together to bond the strips of pith and form a sheet of papyrus. Several attempts have been made to produce papyrus using this methodology, but most have failed to produce papyrus sheets of a quality that has all the apparent characteristics of the best of the excavated material.

Boats and ropes, both produced from harvested papyrus were essential contributors to Egyptian societal development. The earliest papyrus boats gave an essential mobility to occupants of the wetlands, particularly during the annual floods, and the use of papyrus ropes pulled by hundreds of workers allowed Egyptians to move the massive stones required for the building of their monuments as well as other heavy duty tasks. The variety of uses of papyrus harvested by ancient Egyptians testify to its utilisation in modern day terms as a 'sustainable multipurpose crop'. To manage the swamp sustainably the culms were harvested at regular intervals and after cutting a new population developed to replace those that had been removed. However this only achievable while there were clear benefits to the expanding population of Egypt from the use of the harvested papyrus. The collapse in the market for papyrus paper and the transition to wooden boats would have led to a collapse in the demand for the papyrus biomass and at the same time an increasing demand for land on which to grow cereal crops. This led to the clearance of papyrus wetlands and drainage to provide fertile arable land on which to grow cereal crops that supplied a new export opportunity. The demise of the papyrus wetlands was complete as there were no requirements for its ecosystem services in an intensively agricultural environment.

References

Bausch, F., Rosado, M.J., Rencoret, J., Marques, G., Gutiérrez, A., Graf, J., del Río, J. C., Rosenau, T. and Potthast, A. (2022). Papyrus production revisited: Differences between ancient and modern production modes. *Cellulose*, 29, 4931–4950.

Borojevic, K. and Mountain, R. (2011). The ropes of pharaohs: The source of cordage from 'Rope Cave' at Mersa/Wadi Gawasis revisited. *Journal of the American Research Center in Egypt*, 131–141.

Brukner Havelková, P., Dulíková, V., Bejdová, Š., Vacková, J., Velemínský, P. and Bárta, M. (2024). Ancient Egyptian scribes and specific skeletal occupational risk markers (Abusir, Old Kingdom). *Scientific Reports*, 14(1), 13317.

Flieder, F. and Delange, E. (2001). Papyrus: The need for analysis. *Resaurator*, 22, 84–106.

Gaudet, J. (2014) *Papyrus: The plant that changed the world – from ancient Egypt to today's water wars*. New York: Pegasus Books.

Gaudet, J. (2023). *The Pharaoh's treasure: The origin and the rise of Western civilization*. Stroud: Amberley Publishing.

Hassan, F.A. (1997). The dynamics of a riverine civilization: A geoarchaeological perspective on the Nile Valley, Egypt. *World Archaeology*, 29, 51–74.

Jones, M.B. and Muthuri, F.M. (1997). Standing biomass and carbon distribution in a papyrus (*Cyperus papyrus* L.) swamp on Lake Naivasha, Kenya. *Journal of Tropical Ecology*, 13, 347–356.

Leach, B. and Tait, J. (2000). Papyrus. In P.T. Nicholson and I. Shaw (eds), *Ancient Egyptian materials and technology*. Cambridge: Cambridge University Press.

Lewis, N. (1974). *Papyrus in classical antiquity*. London: Oxford University Press.

Mohamed, A. (2023). Methods and materials of making ancient and modern papyrus sheets: A review. *Labyrinth: Fayoum Journal of Science and Interdisciplinary Studies*, 1(2), 74–81.

Muthuri, F.M. and Kinyamario, J.I. (1989). Nutritive value of papyrus (*Cyperus papyrus*, Cyperaceae), a tropical emergent macrophyte. *Economic Botany*, 43, 23–30.

Teeter, E. (1987). Techniques and terminology of rope-making in ancient Egypt. *The Journal of Egyptian Archaeology*, 73(1), 71–77.

Touzeau, A., Amiot, R., Blichert-Toft, J., Flandrois, J.P., Fourel, F., Grossi, V., Martineau, F., Richardin, P. and Lécuyer, C. (2014). Diet of ancient Egyptians inferred from stable isotope systematics. *Journal of Archaeological Science*, 46, 114–124.

Vallejo, I. (2022). *Papyrus: The invention of books in the ancient world*. London: Hodder & Stoughton.

Wallert, A. (1989). The reconstruction of papyrus manufacture: A preliminary investigation. *Studies in Conservation*, 34, 1–8.

Ward, C. (2006). Boat-building and its social context in early Egypt: Interpretations from the First Dynasty boat-grave cemetery at Abydos. *Antiquity*, 80(307), 118–129.

Williams, M. (2019). *Nile Basin: Quaternary geology, geomorphology and prehistoric environments*. Cambridge: Cambridge University Press.

3 The Nile River Basin and Papyrus Wetlands

Introduction

In this chapter we focus on papyrus in the Nile River Basin, which forms the largest of the river basins on the African continent. It is in this catchment that many of the most extensive areas of papyrus wetlands are found. The distribution of the papyrus wetlands in the Nile River Basin is reviewed and the factors that determine their distribution are explored. Evidence from the Nile valley shows how papyrus swamps have had a major impact on the early development of human populations in providing a refuge from annual floods as well shelter in the form of simple housing, boats for transport and a range of artefacts that supported everyday subsistence, as described in Chapter 2. The Lower Nile Valley was one of the cradles of urban civilisations, and was totally dependent on water from the Upper Nile for their survival. These waters have their source in the Upper Nile catchments of the White Nile and the Blue Nile and we highlight how the White Nile is responsible for contributing a relatively stable base flow which is moderated by the movement of water downstream through some of the most extensive papyrus wetlands on the continent, while the Blue Nile, where moderating wetlands are largely absent, is responsible for the annual flooding of the Nile.

Today, as in the past, a majority of inhabitants of the Nile Basin are dependent on the Nile and its headwaters for their supply of water for everyday consumption, as well as supporting their agricultural food production. The supply of this water is threatened by the negative impacts of global climate change while continued and sustainable access to this water is necessary for the survival of millions of people. The requirement for the equitable sharing of the water supply is a source of potential conflict between the counties that occupy the Nile Basin. The history of the search for resolutions to these conflicts will be reviewed and the potential impacts of current man-made climate change will be assessed.

The River Nile and the Nile Basin

The Nile is the longest river in the world at about 6,700 km. It extends over latitudes 4° S to 32° N, with a total catchment area of nearly 3 million km^2,

DOI: 10.4324/9781003131540-3

approximately 10% of the surface area of Africa and covering 11 riparian countries (Tvedt, 2021) (Figure 3.1.). It is estimated that about 250 million people are reliant on the Nile's water in Ethiopia, Uganda, South Sudan, Sudan and Egypt. Nearly all the rainfall that feeds the Lower Nile flowing through Egypt comes from two main tributaries – the Blue Nile and the White Nile. The Blue Nile is fed principally by rainfall in the Ethiopian Highlands and the supply for the White Nile comes mainly from the Upper Nile Basin, found in South Sudan, Uganda, Rwanda and Burundi. This vast catchment contains a remarkable diversity of climate, landscapes and ecological settings. Because it extends over 35 degrees of latitude the Nile Basin can be split into a number of sub-basins each with its own physical, hydrologic, and climatic characteristics (Macklin et al., 2015). The Nile valley also has a long and exceptionally rich history of human occupation. During the Palaeolithic, a period of human prehistory also known as the Old Stone Age, primitive humans used stone and wood tools, learned to build fires, started to keep records and believed in an afterlife by burying the dead. The Nile valley also formed an important corridor for the movement of people during the Palaeolithic and allowed agricultural innovation, social organisation and cultural exchange that led to the emergence of early complex societies.

The countries of the Lower Nile basin, Sudan and Egypt, receive very little rainfall themselves and depend heavily on the Nile for water. The Lower Nile, unlike most other great rivers, flows for much of its length through a very dry desert. Egypt can be divided into three main types of environments – the Nile Delta in the north, the Nile Valley running through the centre and the deserts and oases that flank it. This varied landscape has changed over time impacted by a combination of geological activity, changing climate and human influence.

The current cultivated strip of green, which is the Nile as it flows south to north across the Sahara Desert looks remarkably stable but it is the result of a tempestuous past passing through many phases (Williams, 2019). Said (1993) describes how the basin drainage evolved from a relatively short river. This excavated a canyon up to 4 km deep some 6 million years ago, when the Mediterranean Sea was dry. This river was separated from the rest of Africa by the Nubian massif. The Sudd at that time formed on an enormous lake which covered much of the present Nile Basin within the present-day Sudan and South Sudan. The Equatorial plateau drained towards the west into the Congo/Zaire Basin (Williams, 2019). Further south at the same time the African Rift caused major changes in drainage patterns. Lake Victoria formed within the depression between two arms of the Rift Valley and reversal of the west flowing rivers like the Kagera, the Kafu and the Katonga. The Sudd and Lake Victoria were closed basins until they overflowed to the north about 12,500 years ago, developing the present shape of the Nile Basin.

The route of the modern Lower Nile was established around 10,000 years ago at the onset of the Holocene Epoch Wet Phase. This wet phase produced enough water for branches of the river channel to overflow their

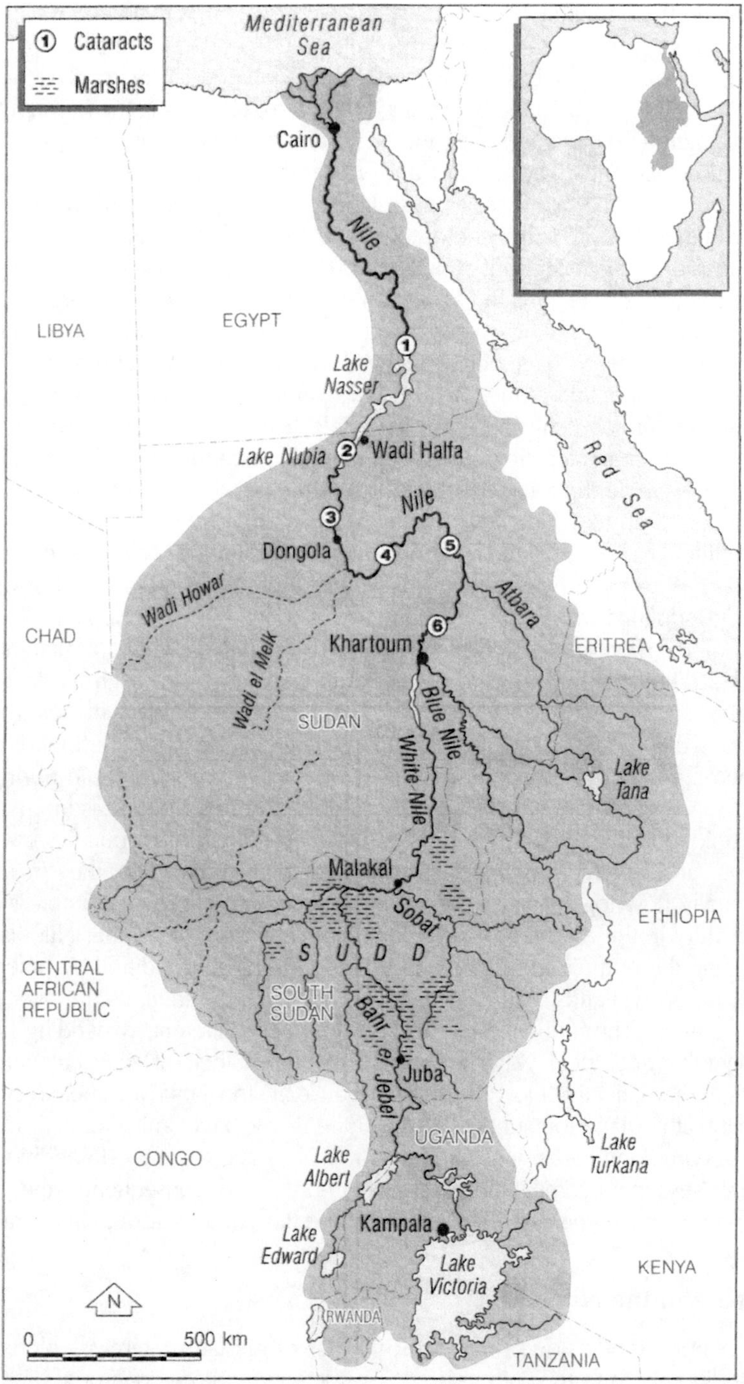

Figure 3.1 Map of the Nile catchment and delta showing location of the Sudd wetlands.

Source: Macklin et al. (2015)

banks and to combine to form one catchment area with a river which had multiple sources (Said, 1993). During the Holocene Wet Phase the entire Sahelian 100 to 500 mm rain belt, which extended along latitude 15° N from the Senegal River to the Nile, shifted 8–10 degrees northwards, expanding the rain-front and increasing the precipitation over a large part of the Sahara. During this time the vegetation became more prolific and wetland extended across much of the landscape. As long as the wet phase lasted the Nile's flow was considerably larger than today. Since the end of the wet phase and the southward shift of the African monsoonal rain front around 5,000 years ago, the flow of the river has tended to decline steadily, but at the same time it has shown very large fluctuations where high flows have dominated. However, in the less frequent periods when very low flows have occurred they have had the most serious impacts on the human inhabitants of the Lower Nile in particular, at times 'causing Egypt to disintegrate into anarchy and dark ages' (Said, 1993, Stanley et al., 2003).

The major characteristic of the Nile channels since their earliest formation is that they have continuously migrated across the floodplain. In some places the floodplain is up to 10 km wide, while in others it narrows to the width of the main Nile Channel, which was cut through sedimentary rocks in the geological past. Where the river cuts through rocks there is no migration, but on the wide floodplains the river can migrate across the whole floodplain. Excavations of the floodplains have shown that the average rate of migration is around 2 km per millennium, but the rates have been as high as 9 km per millennium (Bunbury and Rowe, 2021). Migration of the river means that the area of wetland associated with them also migrates. Because the flow of water in the Nile is highly seasonal, migration rates were probably rapid during the flood season and much reduced or negligible during the dry season.

A unique feature of the Nile is that a number of disparate tributaries have, by geological accident, combined into one long river system after they meet and eventually flow together into the Mediterranean Sea. The overall impression of water flow in the river is of a persistent base flow from the Upper Nile, largely reflecting the outflow from the Lake Victoria basin, but modified by other lake systems and wetlands, particularly the Sudd in South Sudan, as the river flows down-stream. The influence of each reach is very different, depending on the topography and local water balance. The importance of seasonal variation changes dramatically below the Sudd, because the local inflow is the direct input from the Ethiopian tributaries of the Blue Nile. These have limited rainfall seasons and steep upland catchments, and thus relatively fast response. In addition, the rainfall season becomes shorter moving north, and consequently the runoff season becomes progressively shorter and the flow more seasonal in character.

Climate of the Nile Basin

The Nile currently flows through a number of climatic regimes which show large differences from south to north. To the south, in the equatorial highlands of Uganda, Rwanda and Burundi, the climate is hot and wet throughout the year, with two distinctly wetter seasons when the sun is directly overhead. The subtropical Ethiopian highlands have a more seasonal climate regime, with one main wet season during the time of the summer monsoon (June–September) and a shorter wet season earlier in the year (March–April). The Nile's flow regime integrates two parts of the global climate system; the summer monsoon in the northern hemisphere and the equatorial rainfall in the Intertropical Convergent Zone (ITCZ). To the north the rainy season gets shorter and the amount of rainfall decreases. The rainy season in the south extends from April to October but it is confined to July and August in south-central Sudan and further north a desert-type climate exists as far as the Mediterranean. The main characteristics of northern Sudan and Upper Egypt are aridity and a high diurnal temperature range with a daily maximum in June of 47°C.

The annual migration of the ITCZ from the northern to the southern hemisphere and back again follows the zenith of the sun with a 3–4 week lag. The ITCZ is a belt of low pressure generated where the northeast trade winds meet the southeast trade winds. The intensive insolation causes the build-up of mesoscale convective systems forcing moist air upwards and resulting in a band of heavy precipitation. The ITCZ dominates the tropical and sub-tropical climate and is crucial in sustaining tropical ecosystems as it moves depositing as much as 3,000–4,000 mm y^{-1} close to the equator where it can persist all year. Further away, where the ITCZ moves seasonally, precipitation is in the order of 500–1500 mm y^{-1} producing river flood pulses from runoff which results in seasonal patterns of wet and dry in the wetlands. The northward passage of the ITCZ across South Sudan and the southern half of Sudan gives a summer wet season in that region, with mean annual precipitation diminishing northwards from nearly 1,000 mm near Juba on the White Nile to less than 200 mm at the confluence of the Blue Nile and White Nile. The flood pulses that regulate the annual cycle in the Sudd wetlands of South Sudan are traditionally exploited to allow agriculture and annual grazing during the low water periods and fisheries throughout the year. High production rates of grazing areas make wetlands important for flood recession agriculture as well as for grazing of livestock. In these subtropical regions the inter-annual variation in rainfall is large, with possibly several consecutive very dry years that can dry up normally wet-peaty areas and make them vulnerable to oxidation and loss of carbon by microbial respiration and possibly extensive fires. While the ITCZ is the primary regulator of the inter-annual variability of the Nile its flow is also correlated with the movement of the summer monsoon driven by the El Niño Southern Oscillation (ENSO). The ENSO is a recurring climate pattern involving changes in the temperature of waters in the central and eastern tropical Pacific Ocean and associated changes in atmospheric circulation. The Southern Oscillation Index (SOI) is a standardised index based on the sea level pressure differences across the Pacific where negative SOI values coincide with abnormally warm ocean waters across the southern Pacific typical of an El Niño episode. The El Niño is now recognised as a major factor in shaping the climate and climate variability in the Nile Basin. During years when a weak summer monsoon coincides with a strongly negative SOI, the result is severe and prolonged drought and conversely during years of a stronger than average summer monsoon synchronous with a positive SOI, the result is extreme and widespread floods (Williams, 2019).

The subtropical Ethiopian Highlands, which are the source of the Blue Nile, have a more seasonal climate regime, with one main wet season during the time of the summer monsoon (June–September) and a shorter wet season earlier in the year (March–April). The East African summer monsoon receives moisture from the Indian Ocean and brings heavy rainfall to the Ethiopian Highlands between June and September.

Ancient Egyptians appeared to have little knowledge of the hydrology of the Upper Nile region and prior to the discovery of the sources of the Nile less than 200 years ago the causes of the irregular flows were largely a mystery. However, they used gauges called 'nilometers' which were graduated scales cut in the natural rocks in stonewalls to record the swelling of the Nile in the summer which were largely the results of heavy tropical rains in Ethiopia. Nilometers have been found located at Aswan, Karrak and Memphis but the longest quantitative records are from the Roda gauge at the upstream end of Roda Island in Cairo. From March to May the level of the river is at its lowest. In Sudan, the flood begins as early as April but the effect is not felt down river at Aswan until July. The water continues to rise throughout August and September but the maximum is delayed until October at Cairo. When the Blue Nile is in flood it holds back the White Nile and water spills over the river banks, forming an extensive lake and wetlands and delaying the flow of the water. By the time the water reaches Lake Nasser the Blue Nile accounts for almost 70%, the Atbara close to 20% and the White Nile 10%.

Although the records of flood level from the nilometers are detailed in the short term, indicating year to year variations, they are fragmented in the long term so that it is difficult to detect long-term trends but some studies have suggested that there has been increasing aridity over time. Touzeau et al. (2014) made measurements of oxygen isotopes in the mineralised tissues (teeth and bones) of ancient Egyptian mummies to monitor the water they consumed from the Nile during their life. The change in isotope signal indicated that there was a decline in the contribution of the White Nile to the water that reached lower Egypt as a result of a marked decrease in the amount of precipitation at the source of the White Nile that followed a long-term drying trend in the region from around 3500 BCE to 500 CE. The written records of the Egyptian civilisations along with the detailed records of the annual Nile levels have shown the profound impact that the annual Nile floods have had on the every-day lives of Egyptians. For example, the renowned prosperity of Egypt's Ptolemaic era (305–30 BC) was built on cereal exports and ultimately directly tied to the river Nile by the annual summer flooding and irrigation of the land which was of critical support to the region's agriculture. The wealth of Egypt at that time was in large part due to the unrivalled fertility of the Nile, which served as the bread basket to the Ptolemaic kingdom. This era is richly documented and historians have been able to precisely determine the dates of major political events, including revolts against the ruling Ptolemaic family, associated with the failure of the annual floods. As land irrigation was managed through natural flooding, a dramatic reduction in flood water would inevitably jeopardise food supply. This helped trigger economic and political instability and triggered revolts against Ptolemaic rule of Egypt, illustrating how sudden environmental pressures act to destabilise society (Hassan, 1997).

Hydrology of the Nile Basin

Of the three major tributaries that supply water and sediment to the main Nile, the White Nile, the Blue Nile and the Atbara, the White Nile drainage basin covers about half of the total Nile Basin area although it contributes only 32% of the total mean annual discharge and as little as 3% of the Nile's sediment load (Williams, 2019).

The longitudinal profile of the Nile Basin is made up of a series of steep sections interrupted by flatter areas and starts at the source of the Kagera in Burundi at 2,800 m above sea level. The Kagera flows into Lake Victoria at about 1,134 m above sea level after meandering through a series of lakes and swamps, mainly in Rwanda. Other rivers drain the plains of the Serengeti to the southeast of the lake and the swamps of Uganda to the northwest. Before the inception of rifting in Uganda and creation of the western Rift in the Miocene epoch about 8–9 million years ago, the general orientation of the rivers was from the more elevated lands in the east into the present Congo/Zaire Basin lying to the west. The rifting and diversion of drainage later led to the formation of Lake Kyoga and the formation of large lakes, including Lake Albert, on the floor of the Western Rift, as well as Lake Victoria about 0.5 million years ago (Williams, 2019). The drainage connections between Lake Victoria, Lake Kyoga and Lake Albert, were probably established by the Middle Pleistocene with overflow from the Uganda Plateau.

Lake Victoria is the world's largest tropical freshwater lake and the second largest in the world after Lake Superior in North America. Its discharge is regulated by the lake level, which is dependent for 80% of its inflow coming from rainfall over the lake itself. The history of Lake Victoria has seen it dry out completely several times in the past. Historical and geological observations have shown that over the past 100,000 years the lake has completely dried out at least three times (Beverly et al., 2020) and it can dry up relatively quickly, maybe in as little as 500 years. Over that last 60 years the drops in level have probably been caused by a decrease in rainfall but there is also likely to have been falls due to Uganda releasing water to generate hydroelectricity from its dams as the Nile leaves Lake Victoria. Unfortunately, there is conflicting evidence for any long-term trend in the recorded rainfall during recent times (Di Baldassarre et al., 2011) but generally there has been no significant change detected in the annual rainfall in most of the Nile Basin.

The outflow from Lake Victoria to the Victoria Nile river is confined to a single channel, northwards to Lake Kyoga. The outflow from the lake is now controlled by the Kiira and Nalubale dams near Jinja. The Victoria Nile enters the Lake Kyoga swamp complex on its southern shore at 1031 m above sea level some 100 km north of the outfall. The lake is a grass/sedge-filled valley swamp that drains to the northwest towards the Congo. It owes its existence to the upwarping of the western edge of the Lake Victoria basin, which has reduced the gradient and rate of flow in the rivers flowing west, causing 'ponding' and turning them into sluggish swampy tracts (Hughes and Hughes, 1992). The Victoria Nile then flows west through a series of swamps and rapids and then at Murchison Falls forces its way through a gap in the rocks 10 m wide and tumbles 43 m before flowing westward through a swamp into Lake Albert at 620 m above sea level. The Albert Nile or Bahr el Jebel flows northeast to Nimule where it crosses into South Sudan and passes through deep channels and rapids towards Juba and Mongalla. Here the river enters the Sudd swamps. The swamps begin about the latitude of Bor (6°13′ N) and cover the very gently sloping flood plains of three main rivers: the Bahr el Jebel, the Bahr el Zeraf and the Bahr el Ghazal. These flooded areas evaporate large amounts of water so that the outflow of the swamp is only about half of the inflow. The Sudd has a major buffering role on the White Nile and without this influence the White Nile would be a far more seasonal river. At Lake No the Bahr el Jabel turns east and becomes the White Nile. The White Nile is joined from the south by the Sobat wetlands swamps and marshes formed by the spillage of tributaries including the Machar marshes.

The volume of water flowing down the White Nile is largely dependent on the discharge from Lake Victoria. The most significant characteristic of the hydrology of the White Nile is a persistent baseflow, reflecting the out-flow from the Lake Victoria basin, but modified by other lake systems and wetlands as the river flows downstream. The influence of each is very different, dependent on topography and local water balance, which leads to an increase or decrease in the baseflow. The baseflow has also increased in response to rises in the level of Lake Victoria. In 1961/1962 there was a rapid rise in levels of Lake Victoria and at that time the annual flows of water in the White Nile more than doubled at Jinja and similarly at Mongolla. This led to a rapid rise for 3 or 4 years in the outflow from the Sudd although by a much smaller amount than at Mongalla, indicating the damping effect of the Sudd wetlands. The role of wetlands in reducing river flows is a feature of the White Nile basin especially when water spills into the wetlands and subsequently evaporates.

The White Nile provides the Egyptian Nile with a relatively stable base flow which accounts for more than 80% of its low season flow. However, from the point where the White Nile joins with the Blue Nile at Khartoum the seasonal variation in flow is dramatically increased because the local inflow is derived from the tributaries supplied by the Ethiopian Highlands. These areas have short rainfall seasons and steep upland catchments which result in a rapid response in river flows. In addition, the rainfall season becomes shorter northwards so that the runoff season becomes progressively shorter and the flow more seasonal in character. The Blue Nile drains a major part of the Ethiopian highlands where the annual rainfall is confined to a single season and depends on the seasonal fluctuation of the ITCZ. The main Nile below Khartoum flows north through the Sabaloka gorge and is joined after 325 km by its last remaining tributary, the Atabara, which drains the northern portion of the highlands of Ethiopia and also part of Eritrea. Below the Atbara the river flows in a series of wide loops through an arid region of successive cataracts. The Nile enters Egypt below Wadi Halfa where the reservoir formed by the Aswan Dam now extends south of the border. The Aswan High Dam was completed in 1964 but had been proceeded by an initial construction in 1902 and heightened in 1912 and 1933. The role of the reservoir is to provide annual storage and to augment the low flow of the Nile during the irrigation season. As a consequence the Nile hydrology northwards has been extensively modified and is currently controlled by the release of water from the Aswan High Dam.

In terms of sources and sinks for water, the major source for the Nile River is from rainfall around and over Lake Victoria and the headwaters of the Blue Nile in Ethiopia and the headwaters of the Atbara River. Losses are through seepage, evapotranspiration and overbank flows to the swampy lands that fringe the basin in many parts and especially in its equatorial stretch and river discharge from the delta. The major sink areas are located in much of South Sudan, parts of the Sudan and the Nile floodplain and the delta in Egypt. In the case of the Sudd basin alone, close to 50% of the water is lost through the swamps, and considerable quantities are lost even before the water reaches the Sudd itself. Rivers originating in the Ethiopian highlands including the Blue Nile can annually provide more than 80% of the water that flows into the Lower Nile. However, there is a very markedly seasonal flow and is the reason for the very marked seasonal water levels in the Lower Nile which were the source of much concern for the early Egyptian civilisations.

There is still a good deal of uncertainty about several aspects of the hydrology of the Nile Basin (Senay et al., 2014). For example, the average annual discharge from the Nile into the Mediterranean has been variously estimated to be between about 6 $km^3\ y^{-1}$ to 37.5 $km^3\ y^{-1}$. There is also a lack of consensus on the annual contribution of the Blue Nile to the total Nile discharge with estimates ranging from as low as 55% to as high as 84%. Furthermore, the amount of water lost due to evapotranspiration over the Sudd wetland is poorly understood with the most frequent estimates of 50% of White Nile inflows being lost but these estimates have also been as high as 63%. Lack of uniformity in the quality and distribution of weather data and flow observation points in key areas of the Nile Basin have prevented vigorous analysis in determining spatial and temporal distribution of the various water balance components in the basin. Development of remote sensing-based approaches to map the water balance dynamics using satellite derived rainfall, model assimilated evapotranspiration and runoff elements integration enables a higher accuracy of estimates of water balance and provides better contemporary understanding of the Nile Basin hydrology (Rebelo et al., 2012, Pavur and Lakshmi, 2023). The study of Senay et al. (2014), using satellite remote sensing revealed that about 85% of the runoff generated in the equatorial region in the Lake Victoria basin catchment is lost in an area that includes the Sudd wetlands of South Sudan. They also found that the expected average annual Nile flow at Aswan is greater than previously reported (97 km^3 y^{-1} compared to 84 km^3 y^{-1}).

The Nile Basin Wetlands

The Nile Basin wetlands form some of the most extensive wetlands on the African continent. The topography of the African continent, and in particular that of the Nile Basin, with its prevalence of basins generated by tectonic movements, provides the opportunity for massive swamp and marsh development in areas of relatively shallow, slow-flowing waters. These are areas where dense, rapidly growing vegetation, can develop. The distinction between swamps and marshes is that swamps are flat areas covered with vegetation which is flooded to a shallow depth, either permanently or for most of the year, while marshes are areas that are only temporally covered with water when the rivers spill onto the floodplains. As explained previously in Chapter 1, the duration, maximum depth and level range of flooding has been demonstrated to determine the vegetation types and species and conversely, wetland distribution and the presence of key vegetation species are important indicators of the hydrology of the basins. Although the hydrology of the Nile throughout its length has been significantly modified over time by humans, and these modifications have affected the ecology of the river and the land areas adjacent to it, there is a remarkable degree of uniformity of vegetation distribution and species content within the wetlands. In broad terms, although water depth and duration of flooding determine species distribution, all characteristically form closely packed emergent vegetation exhibiting both high rates of continuous growth and rapid microbial activity, leading to rapid decomposition of senescing and dead material. Changes in species distribution can be explained in terms of the relationship between the river level and the consequent patterns of overspill onto the floodplains. For example, in a survey of the Sudd wetlands Howell et al. (1988) found that floating species such as *Cyperus papyrus*, and *Vossia cuspidata* (hippo grass) occupied the deeper water while the shallow-flooded swamp grasses, *Phragmites communis, Echinochloa stagnata, Echinochloa pyramidalis, Typha* spp. and *Oryza longistaminata* were anchored by their rhizomes and root systems to the substrate. The outermost zone, flooded for only a short time each year, was dominated mainly by the densely tufted grass, *Hyparrhenia rufa*. *Vossia cuspidate* also tends to occur at the head of the swamp, due to its ability to withstand strong currents and great fluctuations in water level by remaining rooted and showing rapid stem extension as water levels rise. It also appears to have a high nutrient demand, which is most likely to be met at the inflow end of the swamp.

Because wetlands and lakes cover approximately 10% of the Nile Basin they play an integral role in the hydrology of the area as water spills from the river into the wetlands. Functionally, the wetlands are important in both regulating the flow of the Nile but they also have significant influence on the water chemistry. The effect on water quality is complex as it is dependent on the combined processes of sedimentation, filtration, physical and chemical mobilisation, microbial interactions and uptake by vegetation. The wetlands also provide a range of ecosystem services for the indigenous populations which we will consider later in more detail (Chapter 6). For the earliest herder settlers the wetlands provided a regulated supply of water for cattle which ensured survival during the notoriously erratic dry seasons (Zhao et al., 2022). The subsequent development of farming was also enabled by the supply of water, although increasing populations as agriculture expanded placed enormous pressures on the wetlands, which increasingly needed to be managed to sustain their ecosystem services.

Some of the important features of the most extensive wetland areas in the Nile Basin are described in the following sections.

Nile Delta Wetlands

The Nile Delta in northern Egypt, formed as the Nile enters the Mediterranean Sea, is estimated to have an area of approximately 22,000 km^2 and is one of the largest river deltas in the world (Hughes and Hughes, 1992). It is roughly triangular in shape, approximately 175 km from apex to base and some 263 km wide at the sea. Eight thousand years ago, during the Saharan Neolithic Period, the delta was inundated by the rising sea as the ice caps melted at the end of the last ice age and the Mediterranean Sea rose by 20 m to its present level. The delta then formed as branches of the river flowed into the sea spilling out over the delta depositing thick layers of silt. The delta subsequently consisted of a complex network of river arms, islands, natural levees, alluvial flats, and backswamps where extensive areas of papyrus became established. The Nile Delta was, in the past, known for the large papyrus wetlands and was referred to as the 'papyrus lands' by ancient Egyptians. These areas of swamp were probably used for harvesting of papyrus for the manufacture of paper when there was a market for the product, but when the demand declined it is likely that swamps were cleared for agriculture and this accounts for the widely reported apparent absence of papyrus from the Nile Delta when Napolean invaded in 1798. Interestingly,

Serag (2003) and El-Ghani et al. (2010) have recently recorded the presence of papyrus, where *Phragmites australis–Cyperus papyrus* communities appear to have been spreading in the eastern section of the delta where wetlands persist mostly associated with lakes and lagoons along the seaward face of the delta. These observations suggest that papyrus may not have become extinct in Egypt as suggested, but have survived as small isolated communities in the Nile Delta.

More recently, the completion of the Aswan High Dam in the south of Egypt in 1970 has significantly modified the ecology of the delta system. Before the dam construction recession farming, where crops are established in the flooded areas after the water recedes, had been practised on the floodplain of the delta for over 5000 years. However, since the dam's completion and the end of the end of annual flooding, the area has been cleared of any remaining wetland habitat, and it is farmed year-round using controlled irrigation. The delta is now a very rich agricultural region, accounting for two-thirds of Egypt's agriculture, and is the most populated and industrialised parts of the country. Unfortunately, the dam has also meant that nutrient rich silt that was formerly deposited across the floodplain now settles out in the dam reservoir and silt-free water runs downstream. This has led to the excessive use of fertilisers to compensate for the lack of silt nutrients, and the development of salinity and waterlogging problems probably due to over-irrigation.

Machar Marshes

The Machar Marshes are a large expanse of wetlands found in the eastern part of Sudan and western Ethiopia. They comprise a vast area of swamps and seasonal flood plains interlaced by an intricate reticulate system of water-courses and numerous lakes. The wetland system extends at least 200 km from north to south (between 8°27′ and 9°58′ N) and 180 km from east to west (between 32° 11′ and 34°9′ E) (Hughes and Hughes, 1992). It is estimated the total area of the wetlands are around 9,000 km^2 but this can expand to 20,000 km^2 during the annual rains and river flooding, providing important livestock grazing in the dry season. However, due to their remoteness the Machar Marshes are one of the least studied systems in the Nile Basin, and as a result there is limited information on the seasonal patterns of flooding and the vegetation characteristics. Hughes and Hughes (1992) describe the Marshes as extensive grassy floodplains and permanent swamps dominated either by papyrus along the water courses or by *Phragmites* and *Typha* away from them. The high variability in the timing and intensity of flooding probably means that the areas of permanent swamp dominated by papyrus are limited.

The Sudd Wetlands

The Sudd wetlands of South Sudan form in the floodplain of the Bahr el Jebel river which extends from Lake Albert in northern Uganda to its confluence with the Bahr el Ghazal which flows from the west, where at Lake No the combined rivers become the White Nile which traverses Sudan towards Khartoum. The name derives from an Arabic word meaning obstacle or blockage of river channels. The permanent swamp's width varies from 10 to 40 km and its length is approximately 650 km (Hughes and Hughes, 1992). It is the largest freshwater wetland in the Nile Basin and is one of the largest tropical wetlands in the world. However, estimates vary considerably. According to Rzóska (1976), the wetland had an estimated area of permanent swamp of 30,000–40,000 km^2, and during the height of the rainy season the wetlands could extend over 130,000 km^2 but Rebelo et al. (2012), while noting that area is often uncertain because of the arbitrarily defined boundaries, using satellite remote sensing in 2007/2008, identified a core area of 9,176 km^2 of permanent wetland and the area of seasonal flooding is up to four times larger. There is clearly evidence that there have been very large changes in the area of the swamp over relatively short periods in time, of the order of decades. For example, comparisons of maps of the distribution of the Sudd and its main vegetation types in 1952 and 1980 suggested that the area of permanent swamp in 1952 was about 6,700 km^2 and in 1980 about 19,200 km^2. In this case, this rise in area was largely due to the marked increase in Lake Victoria discharge starting from the early 1960s.

The wetland derives its inflows from Lake Victoria and rainfall over the wetland itself. The extent of the permanent swamp reflects annual riverine inflow closely as this is more variable than direct precipitation. The very high rainfall over Lake Victoria during the early 1960s increased lake outflow considerably and resulted in an increase in the area of permanent swamp after 1964, with a smaller increase in the area of seasonal flooding (Sutcliffe and Parks, 1999). During seasonally high river flows, the channel capacities of rivers are less than the high flow rates and at this time the excess flows leave the river through small channels which pierce the river banks and spill over the alluvial banks to inundate large areas of floodplain on either side of the river. The maximum extent of flooding occurs after the rainy season between October and December. Seasonal inundation drives the hydrologic, geomorphological, and ecological processes and the annual flood pulses are essential to the functioning of the wetland (Rebelo et al., 2012). The 'flood pulse concept' developed by Junk et al. (1989) describes the importance of the hydrological pulse to the functioning of wetlands. It proposes that the annual inundation pattern is primarily responsible for the high productivity of and the biotic interactions within such systems as the Sudd.

The outflow of water after passing through the Sudd wetlands is only about half the inflow on average, although this can vary considerably from season to season. This loss of water is primarily the result of evaporation from open water and transpiration from the wetland vegetation, although some minor losses are due to seepage into the soil. The seasonal floodplain area is greater in the east than the west and it is these that account for the great evaporative losses from the system, since comparatively little water returns from the floodplains to the river as the flood subsides. It has been recognised from the time of the earliest exploration of the Nile that the Sudd's water balance is strongly influenced by these losses, and because of the increasing need for irrigation water in the Lower Nile reducing these losses became a key element in any plans to manage the Nile usage. This has included proposals to bypass parts of the extensive Sudd wetlands to reduce the water losses, which are discussed later in this chapter.

The two most widespread plant species in the Sudd wetland are papyrus and the giant reed, *Typha domingensis* (Denny, 1984). The core area of the permanent swamp is dominated by floating *Cyperus papyrus* along the river margins, but there are also smaller interspersed areas of *Phragmites communis* and *Voissia cuspidata*. Over much of the swamp, papyrus dominates where it is able to form a floating mat and the largest plants at 5–6 m tall, occur close to the flowing river. Large stands of *Typha domingensis* and water hyacinth (*Eichornia crassipes*) occur in shallower water behind the papyrus and the floodplains are dominated by seasonal grasslands (referred to locally as 'toic') of *Oryzia longistaminata* and *Echinocloa pyramidalis*. The *Typha* swamps are most extensive in the central and northern parts of the swamp that are away from the main river channels. They are not floating but are rooted in a firm substrate covered by generally shallow but variable depths of water (Hughes and Hughes, 1992).

In addition to supporting high levels of biodiversity the Sudd provides many other ecosystem services. It is estimated that more than one million people are almost entirely dependent on the wetland, and it forms the backbone of the Nilotic peoples' economy (Rebelo et al., 2012). It is a major source of water for domestic use and for livestock and an important source of fish. Furthermore the floodplain grasslands provide essential dry-season grazing land for livestock. Before the onset of civil wars, which have made researcher access to the region impossible, the number of livestock using the floodplains of the wetland during the dry season was estimated to be 700,000. Papyrus wetlands are also important to the ecology of fish, particularly Tilapia, providing a refuge for the juveniles. Tilapia provides food security to the impoverished population that live close to the wetlands.

Lake Victoria Basin Wetlands

Lake Victoria, the source of the water for the White Nile, has a mean surface area of 69,000 km^2 (Awange et al., 2019) and is the second largest lake in the world. The present Lake Victoria is about 750,000 years old but because of the slowness of the upwarping along the western margin it may have continued to discharge to the west until about 20,000 years ago when its present northern exit at Jinja was established. The lake shoreline of 3,460 km is bordered by Kenya, Tanzania and Uganda with Burundi and Rwanda also lying within the basin (Hughes and Hughes, 1992). A number of important rivers flow into the lake, including the Mara, Kagera, Yala, Nyando, Bukoro and Kalinka. The Kagera River which drains the mountains of Burundi and Rwanda is the furthest and most remote headstream and is considered to be the source of the Nile. The Victoria Nile outflow from Lake Victoria controls the levels of Nile base flow into the Sudd wetlands. Outflow from the lake has since 1956 been regulated by the Nalubaale Dam constructed for hydroelectric power generation at Owen Falls near Jinja. A second dam, Kiira, was later constructed about 1 km from Nalubaale. It was completed in 1999 and began producing hydroelectric power the next year.

The main features of the lake hydrology are the importance of the lake area and the precipitation over it as the water balance of the lake is controlled mainly by precipitation over, and evaporation from the lake. Tributary inflows are small compared with lake rainfall, but are sensitive to periods of high rainfall. The mean rainfall is 1,315 mm yr^{-1} which is 82% of the total inflow, and the remaining 18% is contributed by river inflow. Of the losses, 76% is due to evaporation from the lake, which is relatively constant from year to year, and the remaining 24% is the outflow forming the Victoria Nile at Jinja. Lake levels and outflows are liable to long-term fluctuations, which have been observed in both the historic record since the late nineteenth century and in the prehistoric record preserved in lake sediments. These records show that the lake dried up twice, at 17,000 and 15,000 years ago (Beverly et al., 2020) and in recent times the lake level was high in 1876 but had fallen to almost 4 m below that level by 1900. From that time until 1961 the level of the lake was stable with interannual fluctuations of about 0.5 m, but in 1961 the level rose sharply again to a new record. From 1961 until 1997 the levels were stable again but from 1997 until 2006 there was a year on year decline which was probably due to the combined effects of increased flow from the lake through the new Kirra dam at Jinja, and a regional drought which reduced rainfall (Pavur and Lakshmi, 2023). During this time large areas of fringing papyrus swamps were stranded as the lake margins retreated. There was an un-seasonally high rainfall in November/December 2006 and in February 2007. The lake level then rose by as much as 1 m in the first six months of 2007 and inundated the papyrus again. It has been suggested that the papyrus survived 18 months of exposure by tapping the peat and sediment below the roots and rhizomes at a depth of 0.5–1 m. Increased occurrence of periods of prolonged and heavy rainfall over Lake Victoria since 2007 and in particular after 2020 has led to a continued rise in the level of the lake.

The size of the Lake Victoria Basin catchment area is around 194,000 km^2, supporting a population of more than 40 million people, that is increasing at a rate of 3–3.5% per year. The watershed includes vast wetlands, both within the catchment and along the shores of the lake, many of which are composed of monotypic stands of papyrus. The outer margins of the floating papyrus stands that fringe the lake frequently break away to form large floating islands. The fringing papyrus has an enormous impact on the lake water forming a buffer zone protecting the lake from excess nutrient loading leading to its eutrophication (Ryken et al., 2015). The growing papyrus roots and rhizomes intercept nutrient-loaded sediments and absorb the nutrients, particularly nitrogen and phosphorus. In addition the process of denitrification within the swamp permanently removes elemental nitrogen.

In recent decades the lake has become increasingly eutrophicated due to the intensification of agricultural activity as the population grows around perimeter. The intensification of agriculture is achieved by drainage and encroachment onto the swamps following burning and clearing the papyrus and replacing it with food crops (Ryken et al., 2015). However, the removal of the swamp increases soil erosion by run off and the input of sediment-bound nutrients into the lake. It is estimated that more than 75% of the nitrogen now entering the lake comes from agriculture, and probably 80% of the phosphorus accumulating in the lake comes from municipal and industrial sewage, discharge from urban and agricultural drainage channels and the dumping of untreated sewage from villages and small developments.

The Development of Agriculture in the Lower Nile Region

The Greek historian Herodotus (*c.*485–425 BCE), widely regarded as the father of modern Nile River studies, posed a number of fundamental questions about the source, age, and flood regime of the river Nile. He travelled extensively in Egypt in around 450 BCE, mainly in the delta region and lower Egypt, as well as upstream to the first cataract at Aswan. Herodotus realised that the dark alluvial soils of Egypt were different from the dessert soils of Syria and Libya and inferred that they were derived from the Ethiopian headwaters of the Nile. He suggested that the Egyptian civilisation was 'the gift of the river' as agriculture was only possible as a consequence of a regular hundred days of flooding during the summer months and the annual deposition of silts along the floodplain.

Recent studies have delved into the past by coring through sediments laid down by the Nile (Pennington et al., 2017; Bunbury and Rowe, 2021). This includes drilling programmes over the Nile Delta to look for indicators of early human activity. Every metre goes back about a thousand years and the deposits provide detailed evidence of a changing environment that included the slowing down and stabilisation of the post-glacial sea level rises. Globally, the end of the last ice age and the start of the Holocene epoch approximately 10,000 years ago marked the beginning of a much more favourable climate for human settlement. Weather patterns tended to become more stable and where climate warmed and provided sufficient rainfall for crop growth it stimulated the development of agriculture, practised by settled communities and the evolution of more complex social structures (Pennington et al., 2017).

The Nile Valley, over the early stages of human settlement continued to show significant hydrological variability which was characterised by high to low water levels in the lakes and a contraction of the floodplains in dry conditions and expansion under wetter conditions. Initially, the diverse habitats under humid conditions were ideal for fishing and fowling, supported by the harvesting of wild grains from seasonal floodplains. Macklin et al. (2015) suggested that climatically driven changes in river flow were of critical importance for the development of floodwater farming and they showed that the recorded major disruptions to floodwater farming communities over ancient Egyptian history were principally climatically induced. It has now become clear that river hydrology and floodplain environments in the Nile Valley and delta region has been different in Neolithic, Predynastic, Dynastic, Greco-Roman and medieval times and that these changes have led to large variations in the location and spread of the associated wetlands (Macklin et al. 2015).

Farming and herding in the Nile Valley began around the late sixth millennium BCE. Prior to this the environment of the Nile Valley was rather inhospitable as the river levels were high and the annual summer flood brought by the Ethiopian monsoon made them even higher, ensuring that they were largely inaccessible to any potential settlers. At this time the whole of North Africa experienced a wet climate and farming communities lived across what we now know as the Sahara Desert. However, as the Saharan climate dried, people migrated towards more reliable waters of the River Nile where they occupied the terraces that flanked the marshes of the delta and used the river floods that irrigated the land in the floodplain to grow crops.

During the final stages of the aridification of the Sahara the erosion of the dry valleys added sand and gravel to the Nile Valley, raising the flood plain to form levees, which became suitable for settlement (Pennington et al., 2020). The early settlers found the levees safe from flood waters but close to the marshes for fish and sources of water. These levees were long, raised areas fringing the Nile River and settlements were strung out along them and confined to the tops during floods. During what is referred to as the Old Kingdom, from about 2650 to 2100 BCE, these settlements and towns were restricted to the safe high ground of the levees. Eventually, towards the end of the Old Kingdom the vegetation of the Sahara declined further as droughts became more frequent and wind-blown sand from the Sahara started to drift into the Nile Valley, masking the topography and creating dune fields over the northern settlements (Stanley et al., 2003). The blown sands gradually covered areas further south until by the end of the Middle Kingdom (about 1650 BCE) the whole of Egypt was blanketed with sand. This sand, carried by the trade winds, flowed into the Nile Valley where it was deposited into the river before being carried downstream to form islands in the Nile Delta and also to hasten the consolidation of the delta after its early Holocene inundation. It has recently been shown that these the changes in the topography brought about by the incursion of the sand were accompanied by further dramatic changes as a result of the migration of the river across the delta over the last 5,000 years (Bunbury and Rowe, 2021). It has been estimated that the mean aggregation rate of the delta was around 1 m per millennium while the rate of river migration averaged about 2 km over the same period.

Recent studies have shown that throughout the pharaonic times, episodes of desertification and greening swept across Egypt over periods of about 1,000 years. Cyclic oscillations of rainfall in the Nile Basin caused the Nile flood peaks to rise and fall, as well as periodically drying and then rewetting its margins and the surrounding Sahara (Bunbury and Rowe, 2021). In the wetter times the river in flood carried more silt which was deposited to rebuild its delta and floodplain, and developing into a meandering river. Over time the Nile inhabitants began to understand how the river swelled and abated through its annual cycle and also how the channels and islands behaved over the generations. The increasing knowledge of patterns of inundation led to changes in agricultural practices as well as increasing control of irrigation and management of floodwaters as they arrived with the seasons. As the population grew more land needed to be cultivated so canals were built to carry water to areas where water was stored to extend the growing season. Macklin et al. (2015) suggest that climatically driven changes in river flow are of critical importance for the development of floodwater farming.

During the Predynastic period from around 4400 BCE to 3100 BCE one of the world's first states emerged within the Nile Valley and Delta as people stopped living in smaller communities and came together in cities. Small, regionally diverse centres of culture became transformed into the much larger, single territorial entity of Early Dynastic Egypt. Older deltaic lifestyles that had centred round fishing, gave way to new agricultural practices, settled lifestyles and population growth. Emergent elites then recognised the enormity of the newly agrarian delta's grain surplus and sought to realise it for their own benefit, binding together Upper and Lower Egypt in a single administration and taxed entity as they did so. This 'Primary State' existed for nearly a thousand years before internal factors and external pressures caused its collapse at the end of the Old Kingdom in about 2100 BCE.

During the Early Dynastic period from about 3100 BCE, artificial basin irrigation was developed to benefit from the natural seasonal pattern of Nile discharge by regulating the flow of its high waters to flood the river plain. Arguably, the system of basin irrigation was more sustainable than any other irrigation-based society in human history. The system was set to sustain an advanced civilisation through many political upheavals and climate fluctuations over some 5,000 years. To establish basin irrigation the flood plain was divided into a complex of basins that were fed by canals designed to make the maximum use of the flood water (Said, 1993). The water stayed in the basins for 40 to 60 days, during which the suspended alluvium was deposited, after which it was drained off. By regulating the rates of filling and emptying of the basins the Egyptians were able to irrigate their lands with floods of varying intensity. In this way they were more able to effectively cope with the fluctuations in the annual discharge of the Nile. This basin system of irrigation effectively captured the flood waters for the cultivation of a single winter crop. The main crops were cereals (largely barley and wheat) and flax as they had a growing season which fitted in with the winter climate experienced following the arrival of the floods. The success of this intensification of agriculture lead to the production of sufficient grain surpluses that could be traded for the benefit of the ruling elite who were able to establish a singly administered and taxed entity (Pennington et al., 2017, 2020). Nevertheless Said (1993) considered basin irrigation simply an advanced stage of subsistence agriculture and it was the introduction of the water wheel for lifting water from the irrigation canals to the fields in Ptolemaic times, from 303 BCE to 30 BCE, that allowed for the growth of two crops in each year. Prior to this the basin system of agriculture only allowed for a single winter crop, which was harvested in the spring, leaving the field fallow until the arrival of the next flood. The use of the water wheel also enabled the Ptolemies to greatly expand the agricultural land of Egypt and reclaim a significant part of the Fayum district which until then had formed a lake and an escape for the high floods since Middle Kingdom times. During the Ptolemaic time, Egypt supplied not only an estimated population of 4.9 million with grain but also exported a large surplus to Rome and other cities. Enormous levels of agricultural production were achieved, particularly in the delta and the Fayum region and for a time the Lower Nile region became the breadbasket of the Roman Empire.

A steady decline in population took plâce from the end of the Ptolemaic period and through the Roman rule. The decline accelerated at the end of the third century CE when there was unrest and abandonment of farm land that reduced the cultivated area to almost one half what it used to be during the Ptolemaic period (Said, 1993). Decline in population and land use continued further in the five hundred years between the fourteenth and eighteenth centuries. The once great empire on the Nile was therefore slowly brought to its knees by a century's long drought, consequent economic crises and opportunistic foreign invaders.

By the early nineteenth century the traditional system of basin irrigation was in a state of total neglect and the new ruler at that time, Mohamad Ali, decided to take a new approach which would allow perennial irrigation of the land. The new system was dependent on the building of a barrage, with sluices to allow the suspended silt to pass through during the height of the flood. The barrage was located at the head of the Nile Delta and was high enough to raise the water level of the river to as much as four metres above its normal summer level, from where water was distributed throughout the delta by a series of 'summer' canals (Said, 1993). This development of perennial irrigation allowed the year round use of the land and the cultivation of a cotton crop in particular, that could be sold to generate large amounts of foreign exchange. However, the expansion of perennial agriculture was still limited by the amount of water that was made available through the impoundment behind barrages and it was later decided that much improved seasonal storage could be achieved by damming the river as it entered Egypt at Aswan. The Aswan dam was built in 1902 and its effect was to expand summer agriculture and increase crop production to dimensions hitherto unknown in the history of Egypt (Said, 1993).

Agricultural production and exports from ancient Egypt became the foundation of its wealth. The Nile's annual floods and the fertile soils, formed as result of the deposits of silt, enabled Egypt to build an empire based on its agricultural wealth generated mainly as a result of the export of cereals and papyrus. The relative contribution of these to income from exports must have changed over time as the demand for papyrus first increased and then declined but the yields of cereals increased as flood management for irrigation improved and more land area was cleared for crop production.

The History of Papyrus Swamps in the Lower Nile Region

From the third millennium BCE when the Egyptians discovered that strips of papyrus culm could be used to make sheets of paper for writing on, the papyrus growing in the Nile Valley, and in particular the delta region, was considered to be a 'resource treasured by the kings of Egypt'. This was the beginning of an industry that was to grow to last for more than four thousand years and lead Pliny the Elder, in about 70 BCE, to write in his *Natural History*, 'The nature of papyrus too is to be recounted, for on its use as rolls human civilisation depends, at the most for its life, and certainly for its memory'. Widespread use of papyrus paper can be traced back to the New Kingdom dynastic period around 1550–1070 BCE.

In the early stages of settlement of the Lower Nile the extensive seasonal and permanent wetlands provided a very productive environment where the dispersed human populations could mainly fish for food. However, as the populations increased, settled communities were established and the inhabitants began to drain and cultivate areas of seasonally exposed dry land for annual crops. In the delta region the landscape had gradually evolved from a network of small dynamic streams criss-crossing vast expanses of marshland into wider, more open, well drained floodplains (Bunbury and Rowe, 2021). Nevertheless, substantial areas in the Nile Delta and the Fayum Depression, located about 100 km southwest of Cairo (Williams, 2019), remained that could not be drained for crop production. These were areas of swamp that were permanently flooded as they occupied depressions in the landscape which were refilled by overspill each year as the Nile floodwaters arrived. As we have seen, papyrus is well adapted to these permanently flooded conditions and it probably grew in abundance in these areas as well as other backwaters found along the floodplains. According to Lewis (1974) hardly any village in Egypt could exist without a neighbourhood swamp, indicating the importance of papyrus to the inhabitants. However, they were no longer a major part of the landscape as this was dominated by irrigated agricultural land and even in the Nile Delta and Fayum Depression papyrus swamp was estimated to occupy only respectively 30% and 22% of the potential area.

It is likely that the stands of papyrus were intensively managed and well maintained to produce culms that were homogeneous in size for making high quality papyrus scrolls. It is uncertain if these enterprises were state-owned or private but whichever they were they certainly operated under the close supervision of government agents. Their first priority was to deliver specified quantities of acceptable quality for the government authorities, after which they might produce for their own account but only if they did not compete directly with the state supply.

The first use of papyrus outside of Egypt was by the Greeks in the sixth century BCE but it was probably during the Ptolemaic period (303–30 BCE) that the use of papyrus paper expanded throughout the Roman Empire. Alexander the Great had been welcomed by the Egyptians as a deliverer from the Persians in 332 BCE and following his death Ptolemy took the title of pharaoh and founded a dynasty that ruled Egypt for nearly 300 years until the Roman conquest of 30 BCE. The Ptolemies made Alexandria the intellectual capital of the world, developing the Great Library stocked with hundreds of thousands of papyrus scrolls. There is evidence that after the Roman takeover of Egypt in 30 BCE, there was large scale privatisation of the papyrus swamps and they were managed like any other agricultural land with an increasing focus on the export market. The papyrus was harvested year-round on an almost industrial scale. Papyrus paper production probably reached a peak during the Roman era as societies increasingly developed the skills to communicate through the written word and to pass on these from one generation to the next (Lewis, 1974; Vallejo, 2022).

During the subsequent Byzantine period papyrus exports flourished until 640 CE, after which the Saracens invaded Egypt and established Arab rule. This made it easier for the Arabs to develop and promote rag pulp paper production throughout their empire, eventually leading to a large market for its cheaper product as an alternative to papyrus. As a result, papyrus paper production declined dramatically leading to the drainage of the papyrus swamps in the Lower Nile region and their conversion to grain production. However, the papyrus swamps did not disappear entirely as travellers from medieval times had reported seeing papyrus still growing in Egypt (Leach and Tait, 2000), although by the time of the Napoleonic invasions in 1798 it had apparently almost completely disappeared from Egypt. Agriculturally, demand for cereals for export was stronger than the demand for papyrus, so that the swamp was cleared for crop production.

There are likely a number of reasons why papyrus paper fell out of favour. The main reason is likely to have been the inability to meet the rapidly increasing demand for writing material at an affordable price while competing with cheaper alternatives. Papyrus paper was probably supplying a niche market with a high quality product at a relatively high price. The source material was produced in limited areas of papyrus swamps that needed to be managed carefully to provide uniform papyrus culms sustainably. The manufacture of papyrus paper was, as we have seen in Chapter 2, a skilled and labour intensive process. Its main competitor, initially made from plentiful supplies of rag pulp and then even more plentiful wood pulp, was produced much more easily for a mass market. The process simply involved pounding and stirring rags in water to separate the plant fibres and then collecting the fibres on a mat held horizontally to drain the water away. The precipitous decline in the use of papyrus paper must have impacted on the local population of the delta region and other areas where its manufacture occurred, as the industry at its peak would have been a major employer. However, the increasing demand for cereal production for exports would have increased the demand for agricultural workers and probably compensated for the loss of employment in papyrus paper production. If papyrus was no longer required for the commercial purposes of making paper it was still likely that the swamps would have been cleared to create space for cultivation of cash crops, particularly cereals and other food crops.

Although papyrus swamps were virtually forced into extinction in the Lower Nile as a result of the demand for agricultural land by the end of the eighteenth century, extensive swamps remained largely undisturbed in the Upper Nile. It was these largely impenetrable papyrus swamps that were destined to provide a major constraint on the European exploration of the Upper Nile in the eighteenth and nineteenth centuries.

Papyrus in the Upper Nile

Britain's expansionism in the late nineteenth century involved occupation of the Nile Basin, which started in Egypt and then lead up the Nile to Sudan and Uganda. Britain was primarily concerned with preventing other European powers, particularly Germany and France from muscling in into London's sphere of interest. As a result of this fear during the last ten years of the nineteenth century, Britain occupied and/or annexed Sudan and Uganda. The aims of the British were essentially hydrological and related to Nile control upstream and the benefits of British economic interest in Egypt. The limited amounts of irrigation water available in Egypt, and the abundance of Nile waters waiting to be controlled for the benefits of Egypt and cotton cultivation, made upstream expansion a very rational imperial strategy (Tvedt, 2011). The Upper Nile therefore provided water to allow massive expansion of cotton farms in Egypt. The waters of the White Nile running through one of the biggest wetlands in the world were described as more valuable than gold and Egypt, which had been the granary of the Roman Empire, became the location of the giant cotton farms of the British Empire. These cotton supplies were strategically important as they were exported to mills in Lancashire, England, at a time when supplies from the USA became more difficult and uncertain.

Winston Churchill who had been appointed the British Under Secretary for the Colonies decided in 1907 to visit Africa and undertake a trip from Mombasa on the Kenya coast to Cairo via Uganda and Sudan. From Lake Albert his party followed the White Nile to Khartoum, and Churchill chronicled the journey in a book, *My African Journey*, published in 1908, which one biographer suggested demonstrated the immense energy and assiduity of

Churchill, saying 'it showed how he never relaxed and he had a finger in every pie within his reach'.

The expedition followed the Nile from Lake Albert to Nimule (270 km) but from Nimule to Gondokoro they proceeded by land to avoid the 190 km of 'ceaseless cataract'. Here the Nile becomes navigable and offers an unbroken waterway open to large vessels until the Shabloka cataract, 160 km below Khartoum and 2,400 km from Gondokoro. From Gondokoro northwards Churchill describes two successive 'deserts'; the 'Desert of the Sudd' and the 'Desert of Sand':

> About one hundred miles (160 km) from Gondokoro, the White Nile enters and spills itself in a vast and appalling swamp. Of the action of this tremendous sponge, whether beneficial in regulating the flow, or harmful in wasting the water through evaporation, nothing needs here be said. But its aspect is at once so dismal and so terrifying that to travel through it is a weird experience, for three days and three nights we were continuously in this horrible swamp into which the whole of the United Kingdom could be easily packed.

He then adds:

> The papyrus plant is in itself a beautiful, graceful, and venerable thing. To travel through the Sudd, is to hate it for evermore. Rising fifteen feet [4.5 m] above the level of the water, stretching its roots twenty or even thirty feet below, and so matted and tangled together that elephants can walk safely upon its springy surface, papyrus is the beginning and the end of this melancholy world.

Despite his discomfort, Churchill, living up to his reputation as a dedicated colonialist, took every opportunity to shoot game. He even suggested that back home in the UK there were 'unenterprising folk' who spend fortunes each year in shooting, on home soil, game from well-counted herds of more or less tame deer, 'when in a month, and for less expense than the years rent of their forest, they could pursue and shoot wild animals of every kind in their natural haunts and gain experiences that would last them all their lives'.

Tarje Tvedt, in his book on the history of the Nile (Tvedt, 2021), refers to what became called the 'Jonglei Canal' as a new aqueduct to the north. The idea for a canal to by-pass much of the Sudd was first proposed by Sir William Garstin, Undersecretary of the State of Public Works of Egypt in a British Government report published in 1907. It was proposed to cut a canal 360 km long that would hasten the river current through the Sudd and thus increase the amount of water reaching the cotton fields of Egypt. There then followed a long period between 1919 and 1954 in which as many as nine comprehensive plans were drawn up to overcome a combination of technical, economic and political problems, but none were implemented. Then in 1954 and 1955 the so called 'Jonglei Investigation Team' published a five-volume report on the consequences of the Jonglei Canal for the local ecology and economy that Tarje Tvedt (2021) considered to be one of the best research reports on a development project that has ever been produced in Africa. The report concluded that the project could not be implemented largely because it would be detrimental to the needs and interests of the local population. Following Sudan's independence in 1956 the proposal was put into abeyance, but in the early 1980s it was taken up again jointly by Egypt and Sudan as the 'Jonglei Development Project' and the excavation of the canal was started. However, by 1993 the project had collapsed after local objections and the killing of some of the Pakistani workers employed by the French construction company. Tarje Tvedt (2021) concluded that 'The politicians have come and gone, and countries have come and gone, but the Nile ecology and Nile hydrology remains the same'. Meanwhile, negotiations about the resumption of the Jonglei Project and implementation continue.

Continuing civil disturbances in the area of the Jonglei Canal have made impossible long term research which can weigh in the balance the benefits of conveying much needed water for irrigation downstream against the advantages and disadvantages of the project to the people of the area through which the canal passes. However, short-term studies, which were possible between 1976 and 1983, have been reported in *The Jonglei Canal: Impact and Opportunity* by Howell et al. (1988), which the editors described as 'an opportunity to present the scientific facts ... together with an objective interpretation to provide a greater comprehension of the local issues and the way in which they might be resolved'.

Access to the Nile's Water

Throughout human occupation of the Nile Valley, survival has depended on the annual inundation of the Nile which over time has been increasingly regulated by an elaborate system of canals and dams to distribute the floods to the fields, renewing the soil with fresh deposits of silt to maintain productivity of crops. Until the beginning of the twentieth century, Egypt was the only basin state that had made use of the waters of the Nile. Until then Egypt's right to make exclusive use of that water had not been challenged and the question of the division of the waters of the Nile among the basin states had never been raised. This is despite the fact that the rivers originating in the Ethiopian highlands provide more than 85% of the water that flows into the Nile River and the rest of the Nile's waters come from the White Nile, which flows from South Sudan, Uganda, Rwanda and Burundi. The first serious scientific studies of the hydrology of the Upper Nile were published in 1904, suggesting that water storage to regulate for supply to Egypt could be achieved by construction of dams along the Blue Nile and Atbara and at the same time preliminary studies of the upper White Nile basin suggested that losses as the water flowed through the Sudd swamps could be reduced by improving flow in the natural channels by dredging and by straightening them and raising their banks. The plans for controlling and storing the waters of the Nile were primarily directed to supplying the largest part of the waters of the entire Nile Basin to Egypt and since the 1920s Egypt has used colonial-era treaties to claim near-total control of the Nile. The 1929 Anglo-Egyptian Treaty estimated the average annual flow of the Nile as 84 billion cubic metres. It allocated 48 billion cubic metres to Egypt and four to Sudan. The 1959 Nile treaty increased Egypt's portion to 55.5 billion cubic metres (66% of Nile's water) and Sudan to 18.5 billion cubic metres (22%). The rest, 10 billion cubic metres (or 12% of the Nile's water), was set aside to account for seepage and evaporation. Despite Ethiopia's massive contribution to the Nile's flow, its rights were not recognised and they were not allocated any of the Nile water.

However, in 2011 Ethiopia announced its decision to build the Grand Ethiopian Renaissance Dam (GERD) on the Blue Nile. The dam is about 40 km east of Sudan. Ethiopia has argued that the dam will benefit the downstream countries by smoothing variations in the water flow and increasing water availability during low flow months. It is also argued that the dam will reduce damage from floods and Sudan will be able to purchase electricity generated by the dam. The immediate issue is that water flow to the downstream countries might be negatively affected during the filling of the dam's reservoir (Wheeler et al., 2016). Negotiations between Ethiopia, Sudan and Egypt have failed to produce an agreement. The dam is expected to produce 750 megawatts of electricity and as 60% of Ethiopians do not have access to electricity the dam has the potential to transform development in the country. The dam has been estimated to cost around US$5 billion on completion. Although their ability to access international finances was apparently frustrated by Egypt, the Ethiopians used crowdfunding to finance the project and they also received a US$1 billion grant from the Exim Bank of China.

A critically important requirement is for the basin states to create an equitable water allocation policy that not only addresses 'historical rights' such as Egypt's long-held claims to two-thirds of the Nile flow, but also to the needs of upstream countries like South Sudan and Ethiopia, which require water to grow their economies. In the future countries in the Nile Basin will need to work more closely together to ensure the best use of the region's resources. There will be a need to trap and store rainfall in the wet, flood risk years, and equitably distribute the water in the dry years. Failure to work together to prepare could have catastrophic consequences.

Clearly, there need for agreement based on accurate data that govern the allocation of water resources in the Nile Basin. In particular, there is a serious lack of data on how much water there is because the level of uncertainty from the ground observations. Unfortunately, the number of functioning hydrological monitoring stations have significantly declined over the last 50 years due to a combination of lack of funds and a lack of expertise to run them, often in remote locations. However, there have been a number of local measurements of rainfall, runoff and evapotranspiration, although they have not generally been integrated spatially across the entire Nile Basin. Major advances in monitoring hydrology since the turn of the century has been made following the launch of hydrological satellite observation missions, the most significant of which has been NASA's satellite mission named the Gravity Recovery and Climate Experiment (GRACE) (Hasan et al., 2021), which makes repeated assessments of total changes in ground water volume within its field of view. The data are available on a monthly basis and have been used to determine the total available water storage in the basin between 2002 and 2020 and this has shown that the total available water storage could reach on average about twice the currently estimated storage of 88 billion cubic metres per year. This data may in the future contribute to the discussion of how much water is allocated in the basin water sharing agreement.

Conclusion

The Nile Basin is the largest of the river basins on the African continent. Three major tributary basins supply water and sediment to the main Nile: the White Nile, the Blue Nile, and the Atbara River. Of these the White Nile drainage basin covers just over half of the total Nile Basin but despite its vast catchment area the White Nile provides only approximately one third of the total mean annual discharge and an even smaller contribution to the sediment load. It does however maintain perennial flow in the Nile during drought years in Ethiopia where the Bue Nile rises. Shallow sub-basins along the course of the White Nile with slow flowing waters allowed the development of extensive wetlands in the Upper Nile before it joined the Blue Nile on its route to lower Egypt and the Nile Delta. Silts deposited from the river accumulated at the delta region slowed the flow of the river to allow further development of extensive wetlands. These were conditions in which vast areas of papyrus dominated wetlands developed in the Lower Nile region. The hydrologic history of the Nile is one of repeated wet and dry phases and times of high and low flow at time scales varying from millennial to centennial to decadal. The Nile Basin was a home for prehistoric human groups for more than a million years and it was during the Predynastic period from around 4400 BCE to 3100 BCE that one of the world's first states emerged within the Nile Valley and Delta as people stopped living in smaller communities and came together in cities.

In the early stages of settlement of the Lower Nile the extensive seasonal and permanent papyrus wetlands provided a very productive environment where the dispersed human populations could mainly fish for food. However, as the populations increased, settled communities were established and the inhabitants began to drain and cultivate areas of seasonally exposed dry land for annual crops. Papyrus wetlands were no longer a major part of the landscape as this was dominated by irrigated agricultural land but it is likely that remaining stands of papyrus were intensively managed and well maintained to produce culms for making high quality papyrus scrolls. The papyrus was harvested year-round on an almost industrial scale, increasingly for export. Papyrus paper production probably reached a peak during the Roman era as societies increasingly developed the skills to communicate through the written word and to pass on these from one generation to the next. After the Saracens invaded Egypt and established Arab rule in 640 CE, papyrus paper production declined dramatically due to competition from cheaper rag paper and this led to the drainage of the papyrus swamps in the Lower Nile region and their conversion to grain production. At this point, papyrus wetlands largely disappeared in lower Egypt.

References

Awange, J.L., Saleem, A., Sukhadiya, R.M., Ouma, Y.O. and Kexiang, H. (2019). Physical dynamics of Lake Victoria over the past 34 years (1984–2018): Is the lake dying? *Science of the Total Environment*, 658, 199–218.

Beverly, E.J., White, J.D., Peppe, D.J., Faith, J.T., Blegen, N. and Tryon, C.A. (2020). Rapid Pleistocene desiccation and the future of Africa's Lake Victoria. *Earth and Planetary Science Letters*, 530, 115883.

Bunbury, J. and Rowe, R. (2021). *The Nile: Mobility and management*. Cambridge: Cambridge University Press.

Denny, P. (1984). Permanent swamp vegetation of the Upper Nile. *Hydrobiologia*, 110, 79–90.

Di Baldassarre, G., Elshamy, M., van Griensven, A., Soliman, E., Kigobe, M., Ndomba, P., Mutemi, J., Mutua, F., Moges, S., Xuan, Y., Solomatine, D. and Uhlenbrook, S. (2011). Future hydrology and climate in the River Nile basin: A review. *Hydrological Sciences Journal*, 56, 199–211.

El-Ghani, M.A., El-Fiky, A. M., Soliman, A. and Khattab, A. (2010). Environmental relationships of aquatic vegetation in the fresh water ecosystem of the Nile Delta, Egypt. *African Journal of Ecology*, 49, 103–118.

Hasan, E., Tarhule, A. and Kirstetter, P.E. (2021). Twentieth and twenty-first century water storage changes in the Nile River basin from GRACE/GRACE-FO and Modelling. *Remote Sensing*, 13, 953.doi:10.3390/rs13050953.

Hassan, F.A. (1997). The dynamics of a riverine civilization: A geoarchaeological perspective on the Nile Valley, Egypt. *World Archaeology*, 29, 51–74.

Howell, P., Lock, M. and Cobbs, S. (eds). (1988). *The Jonglei Canal: Impact and opportunity*. Cambridge: Cambridge University Press.

Hughes, R.H. and Hughes, J.S. (1992). *A directory of African wetlands*. Gland: IUCN.

Junk, W.J., Bayley, P.B. and Sparks, R.E. (1989). The flood pulse concept in river-floodplain systems. *Canadian Special Publication of Fisheries and Aquatic Sciences*, 106, 110–127.

Leach, B. and Tait, J. (2000). Papyrus. In P.T. Nicholson and I. Shaw (eds), *Ancient Egyptian materials and technology*. Cambridge: Cambridge University Press.

Lewis, N. (1974). *Papyrus in classical antiquity*. London: Oxford University Press.

Macklin, M.G., Toonen, W.H., Woodward, J.C., Williams, M.A., Flaux, C., Marriner, N., Nicoll, K., Verstraeten, G, Spencer, N. and Welsby, D. (2015). A new model of river dynamics, hydroclimatic change and human settlement in the Nile Valley derived from meta-analysis of the Holocene fluvial archive. *Quaternary Science Reviews*, 130, 109–123.

Pavur, G. and Lakshmi, V. (2023). Observing the recent floods and drought in the Lake Victoria Basin using Earth observations and hydrological anomalies. *Journal of Hydrology: Regional Studies*, 46, 101347.

Pennington, B.T., Sturt, F., Wilson, P., Rowland, J. and Brown, A.G. (2017). The fluvial evolution of the Holocene Nile Delta. *Quaternary Science Reviews*, 170, 212–231.

Pennington, B.T., Wilson, P., Sturt, F. and Brown, A.G. (2020). Landscape change in the Nile Delta during the fourth millennium BC: A new perspective on the Egyptian Predynastic and Protodynastic periods. *World Archaeology*, 52(4), 550–565. doi:10.1080/00438243.2020.1864463.

Rebelo, L-M., Senay, G.B. and McCartney, M.P. (2012). Flood pulsing in the Sudd wetland: Analysis of seasonal variations in inundation and evaporation in South Sudan. *Earth Interactions*, 16, 1–19.

Ryken, N., Vanmaercke, M., Wanyama, J., Isabirye, M., Vanonckelen, S., Deckers, J. and Poesen, J. (2015). Impact of papyrus wetland encroachment on spatial and temporal variabilities of stream flow and sediment export from wet tropical catchments. *Science of the Total Environment*, 511, 756–766.

Rzóska, J. (ed.). (1976). *The Nile: Biology of an ancient river*. The Hague: Springer Science & Business Media.

Said, R. (1993). *The River Nile: Geology, hydrology and utilization*. Oxford: Pergamon.

Senay, G.B., Velpuri, N.M., Bohms, S., Demissie, Y. and Gebremichael, M. (2014). Understanding the hydrologic sources and sinks in the Nile Basin using multisource climate and remote sensing data sets. *Water Resources Research*, 50, 8625–8650.

Serag, M.S. (2003). Ecology and biomass production of *Cyperus papyrus* L. on the Nile bank at Damietta, Egypt. *Journal of Mediterranean Ecology*, 4, 15–24.

Stanley, J.D., Krom, M.D., Cliff, R.A. and Woodward, J.C. (2003). Nile flow failure at the end of the Old Kingdom, Egypt: Strontium isotopic and petrologic evidence. *Geoarchaeology*, 18, 395–402.

Sutcliffe, J.V. and Parks, Y.P. (1999). *The hydrology of the Nile*. IAHS Special Publication no. 5. Wallingford: International Association of Hydrological Sciences Press.

Touzeau, A., Amiot, R., Blichert-Toft, J., Flandrois, J.P., Fourel, F., Grossi, V., Martineau, F., Richardin, P. and Lécuyer, C. (2014). Diet of ancient Egyptians inferred from stable isotope systematics. *Journal of Archaeological Science*, 46, 114–124.

Tvedt, T. (2011). Hydrology and empire: The Nile, water imperialism and the partition of Africa. *The Journal of Imperial and Commonwealth History*, 29, 173–194.

Tvedt, T. (2021). *The Nile: History's greatest river*. London: I.B. Tauris.

Vallejo, I. (2022). *Papyrus: The invention of books in the ancient world*. London: Hodder & Stoughton.

Wheeler, K.G., Basheer, M., Mekonnen, Z.T., Eltoum, S.O., Mersha, A., Abdo, G.M., Zagona, E.A., Hall, J.W. and Dadson, S.J. (2016). Cooperative filling approaches for the Grand Ethiopian Renaissance Dam. *Water International*, 41, 611–634.

Williams, M. (2019). *Nile Basin: Quaternary geology, geomorphology and prehistoric environments*. Cambridge: Cambridge University Press.

Zhao, X., Liu, Y., Thomas, I., Salem, A., Wang, W., Alassal, S.E., Jiang, F., Sun, Q., Chen, J., Finlayson, B., Wilson, P. and Chen, Z. (2022). Herding then farming in the Nile Delta. *Communications Earth & Environment*, 3, 88. doi:10.1038/s43247-022-00416-7.

4 Carbon and Water Cycling in Papyrus Wetlands

Introduction

This chapter reviews the continuous movements or fluxes of carbon in the form of carbon dioxide or methane and water between papyrus vegetation and the atmosphere, and their contribution to the functioning of the wetland ecosystems. These processes are cyclic, and the process for carbon includes assimilation of the carbon from the atmosphere in photosynthesis, its distribution in the papyrus vegetation and the eventual return to the atmosphere by the process of respiration as the vegetation dies and decomposes. The cycling of water in the hydrological cycle is a facilitator of the carbon cycle. Increasing knowledge of these processes in papyrus wetlands is being achieved through measurements of gas exchange at scales ranging from the leaf, through the canopy and to whole ecosystems using land based and satellite located sensors. The core of the chapter is a review of papyrus wetlands as carbon sinks and sources of the greenhouse gases carbon dioxide and methane and the influence of the hydrological cycle in papyrus wetlands on their greenhouse gas budgets. The cycling of water between the atmosphere, vegetation, and back again, drives the local climate. Rainfall in the river headwaters drives the seasonal cycle of river flooding and in turn influences the functioning of papyrus of wetlands. The chapter concludes with an overview of the impacts of future variability of weather and climate under projected climate change scenarios on carbon cycle in papyrus wetlands.

The Carbon Cycle in Wetlands

In order for the living components of ecosystems to survive all the chemical elements that make up the living system must be cycled continuously in what we refer to as biogeochemical cycles. The term biogeochemical refers to the biological, geological and chemical aspects of each of the cycles. The biotic compartment is the biosphere and the abiotic compartments are the atmosphere, hydrosphere and lithosphere. Biogeochemical cycles are the pathways by which chemical compounds cycle the biotic and the abiotic

DOI: 10.4324/9781003131540-4

compartments of the earth (Schlesinger, 1997). It is through these biogeochemical cycles that the component elements of the plant, mainly in different forms of carbon, nitrogen, phosphorus and sulphur are used in the construction of the plant and then released after death and decomposition. The globally most significant biogeochemical cycle is the carbon cycle as the cycles of other nutrients, including nitrogen, phosphorus and sulphur are more localised.

In the carbon cycle the process of photosynthesis in vegetation converts atmospheric CO_2 to organic carbon in the form of simple sugars initially and then more complex materials including cellulose and lignin, the key structural components of plants. During this metabolism the process of respiration returns some of the CO_2 back to the atmosphere and following the death of the vegetation decomposition by microorganisms returns more. Carbon that is not released is converted to stable forms resistant to decomposition either associated with soil aggregates or under anaerobic conditions to form peat. This sequestered carbon can remain stored below ground for long periods of time but on combustion it will release CO_2 back to the atmosphere and so complete the full carbon cycle. The carbon cycle therefore, has two interconnected sub-cycles. In the 'rapid' cycle the carbon stored in living vegetation is released rapidly through respiration and decomposition by microorganisms following the senescence and death of the vegetation. However under particular sets of circumstances a variable proportion of this carbon can remain out of circulation for longer periods of time in the 'slow' cycle.

Ecosystem structure and composition have strong influences on the functioning of the biogeochemical cycles. In the case of wetlands, the continuous presence of water around the roots and rhizomes results in an anoxic environment that has an overriding impact on the carbon metabolism (Schlesinger, 1997) (Figure 4.1). This is because oxygen is only sparingly soluble in water and it diffuses about 10^4 times more slowly in water than in the air. Consequently, within a few millimetres of water depth, the environment changes from aerobic and becomes increasingly anaerobic as oxygen levels are depleted. This leads to change in the metabolism of the roots and rhizomes which is characterised by very low rates of respiration and decomposition of dead material by microorganisms that are using various forms of anaerobic metabolism. Biogeochemical cycling in freshwater wetlands is therefore controlled by the reduction, or redox, potential of the environment and by microbial transformations of nutrient elements that occur under conditions in which oxygen levels are depleted. Redox potential is the term used to express the tendency of an environment to receive or supply electrons. High redox potentials occur in environment where oxygen is freely available as an electron acceptor. All redox reactions can be broken down into two different processes, a reduction process and an oxidation process. In respiration, oxidation of an electron donor yields electrons for reduction of an external electron acceptor. Different microorganisms may

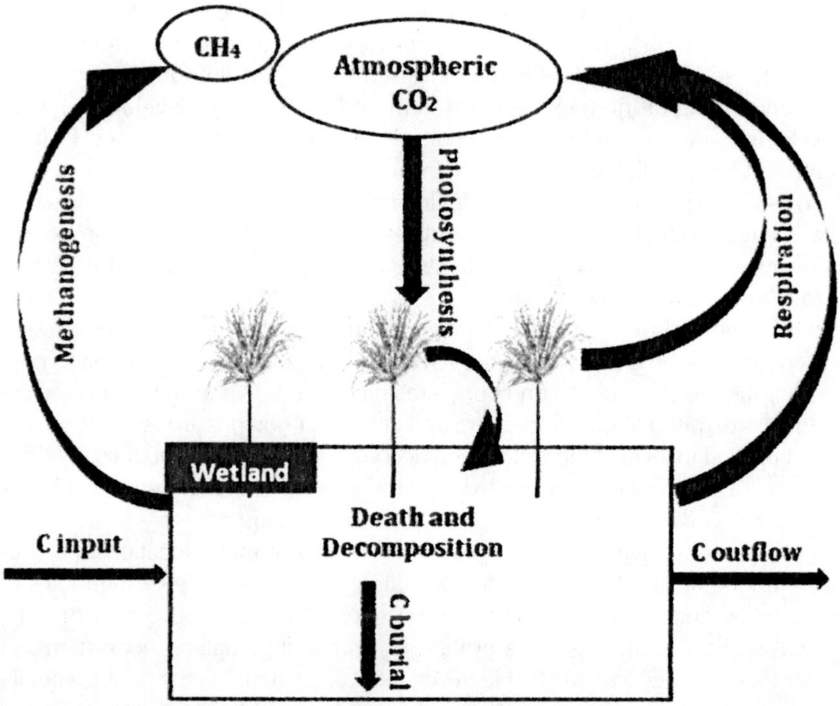

Figure 4.1 Simplified schematic overview of carbon cycling by photosynthesis, respiration and methanogenesis in wetlands.
Source: Were et al. (2019)

utilise different electron acceptors including oxygen, nitrate/nitrite, manganese oxide, ferric iron and sulphate. Anaerobic microbial processes, primarily denitrification, sulphate reduction and methanogenesis are responsible for the release of nitrogen, hydrogen sulphide and methane from the wetlands (Candry et al., 2023).

The Carbon Cycle in Papyrus Wetlands

As we have seen, the papyrus plant is one of largest recorded herbaceous or non-woody species, with culms growing to a height of more than 5 m and an above water standing biomass of about 30 t C ha^{-1} (Chapter 1). The culms are topped by large spherically shaped reproductive umbels that produce masses of minute seeds when they flower but also function as the main photosynthetic surface. The umbel is typically about 50 cm in diameter and consists of several hundred cylindrical rays, each of which extends into three to five flattened, leaf-like structures, which are botanically referred to as bracteoles. As we have seen, papyrus uses the C_4 photosynthetic pathway and although at first sight the wetland environment appears to be an unlikely habitat for C_4 species, which are generally assumed to be best adapted to hot and dry conditions, the high efficiencies of solar radiation and nitrogen use characteristic of C_4 species is hypothesised to have an important impact on the functioning of the papyrus ecosystem and its ability to develop as highly productive and competitive monocultures in environments exposed to high light and low available nitrogen environments (Jones, 1986).

In established papyrus swamps the floating vegetation frequently overlies several metres depth of detritus, sludge and peat. The peat has formed in the oxygen depleted environment below the floating papyrus rhizome mats. The detritus forms from the dead and decomposing papyrus vegetation that collapses below the rhizome and over time this detritus continues to form more highly decomposed peat. As the peat accumulates it forms a 'sink' for carbon that remains only as long as the anaerobic conditions are maintained. When, for example, the detritus is exposed to the atmosphere and aerobic conditions are established as a result of a fall in level of the water, then decomposition is rapid and these systems become net sources of carbon dioxide to the atmosphere because carbon losses in respiration will exceed carbon gains in canopy photosynthesis.

From measurements made on papyrus wetlands at many locations, particularly in East Africa, it has been demonstrated that they have very high rates of primary productivity. In fact, the highest rates of close to 30 g DM m^{-2} d^{-1} probably place it among some of the most productive natural ecosystems on earth (Saunders et al., 2014). However, not all papyrus wetlands are equally productive and lower values are likely due to site-specific differences associated with limitations posed by climate, or by hydrological conditions such as variable water depth and nutrient availability. Invariably, the highest values of primary productivity are associated with floating swamps, which provide a very stable environment throughout the year. In these conditions plant development and growth is a stable and continuous year-round process, with individual culms emerging sympodially from the plant rhizomes. Sympodial growth is a branching pattern where the terminal bud ceases to develop and growth is continued from lateral buds, forming clumps of rhizomes. This means that at any one time for a given area the plant culms are present in multiple age classes. They appear, mature and senesce in a continuous process of growth. Senescence is followed by death and decomposition and dying culms are continuously replaced. The result is that the typical culm in a productive swamp has a complete life cycle of between six months and a year. The plant canopy appears to be unchanged and this ensures that more than 90% of incident solar radiation is intercepted by fully active photosynthetic vegetation, ensuring maximum possible solar energy utilisation (Jones and Muthuri, 1985). Furthermore, the fact that papyrus utilises the C4 pathway of photosynthesis ensures the high rates of CO_2 assimilation in photosynthesis (Jones, 1987) which translate into large carbon stocks held in both the aerial and submerged plant material. In fact, the combined aerial and submerged biomass can be as high as 88 t C ha^{-1}, which is comparable to forest ecosystems that are widely regarded as the largest, land-based, reserves of carbon (Muthuri et al., 1989). In addition, significant detrital and peat-like deposits form within the papyrus wetlands where anaerobic conditions prevail, which severely limit the oxidation and decomposition of organic matter (Were et al., 2019). These deposits accumulate predominantly on the bottom of the waterway or lake where detritus rains down from the floating papyrus above. The carbon stored in these deposits can be many times greater than that found in the living biomass. If the carbon stored in the detrital and peat deposits are included in the calculation of the total carbon stock associated with the papyrus wetland, depending on the depth of the peat, the estimated amount is in excess of 700 t C ha^{-1}. This is a value that far exceeds the measured carbon stored in tropical forest ecosystems (Saunders et al., 2014).

The cycling of carbon in the papyrus ecosystem can be considered to start with the assimilation of CO_2 from the atmosphere in the process of photosynthesis; the process that transforms carbon from its oxidised form as CO_2 in the atmosphere to a reduced organic form that ultimately results in plant growth. As we have seen, the rate of photosynthesis in papyrus are high and this results in a high net primary productivity of the papyrus vegetation (Figure 4.2). However, the carbon cycling also involves the return of CO_2 to

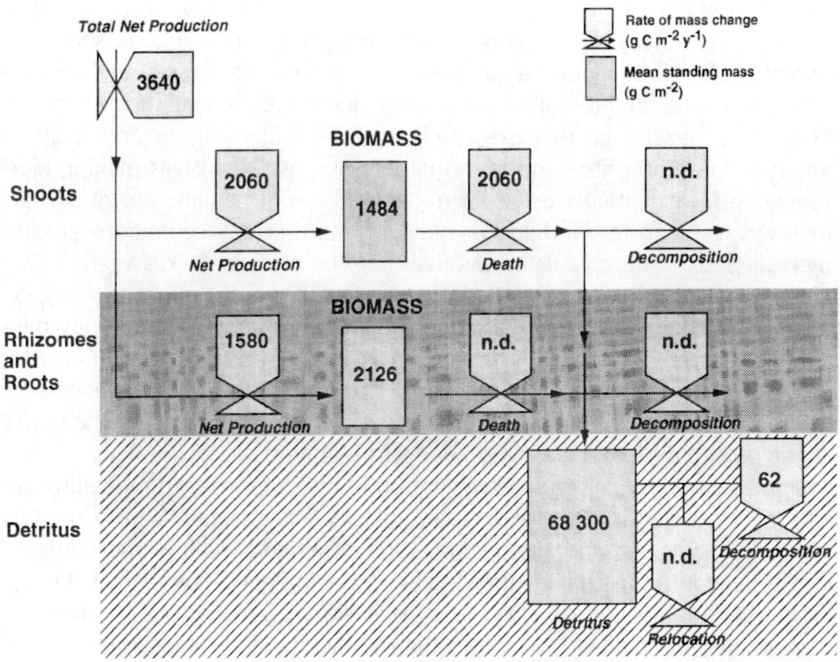

Figure 4.2 Distribution of carbon and estimates of carbon fluxes in a papyrus swamp on Lake Naivasha, Kenya. The arrowed boxes show the rate of change in carbon (gC m^{-2} y^{-1}) and the rectangles show the mean mass (quantity) of carbon (gC m^{-2}).

Source: Jones and Muthuri (1997)

the atmosphere resulting from plant respiration and decomposition of dead organic material, including the relatively short-lived culms and umbels. In wetlands in the tropics the rates of decomposition are high due to consistently warm temperatures and moist conditions but in wetlands the dead material that makes its way below the surface of the water experiences increasingly low oxygen (anoxic) conditions where the rate of decomposition and carbon dioxide release is massively reduced. This leads to long term storage of the dead material as peat deposits that lock up carbon as long as they remain undisturbed. Disturbance most frequently occurs in the form of drainage, natural falls in water levels or where dead material is swept away by water movement. These forms of disturbance can expose the peat to increased atmospheric oxygen concentrations where aerobic decomposition rapidly releases carbon dioxide back the atmosphere.

Measurements of the fluxes of carbon dioxide between the vegetation and the atmosphere enables the calculation of the balance between carbon gains in photosynthesis and losses in respiration. When this is done a net gain in carbon by the vegetation is considered to be sequestered carbon and in the long term the amount of carbon sequestered is the accumulated organic matter that in wetlands forms the peat deposits. The direct measurement of the amount of accumulated peat on an annual basis is very difficult because, as a proportion of the peat that has already accumulated over years, it is small. For short term estimates of carbon sequestration we rely on measuring directly the fluxes of CO_2 using a technique called eddy covariance. Eddy covariance is used to measure the exchange of gases, such as CO_2, water vapour and methane between the atmosphere and terrestrial ecosystems. It involves making high-frequency measurements using instruments like sonic anemometers and gas analysers to measure wind speed, direction, and gas concentrations at rapid intervals. Fluxes are calculated by analysing the covariance between vertical wind velocity and gas concentrations to estimate the net exchange of gases between the ecosystem and the atmosphere. Summed hourly and daily measurements of net CO_2 exchange can provide annual estimates of net carbon accumulation in the wetland ecosystem. The development of this technology and the establishment of monitoring networks have revolutionised the understanding of ecosystem scale carbon budgets. Furthermore, increasingly this data is supplemented by linkages between ground measurements to modelling, the use of remote sensing and geographic information systems (GIS) to produce scaled estimates of landscape, regional and continental scale carbon budgets. This collection of flux data is directed at answering questions such as: Do tropical wetlands operate as effective carbon sinks and can they be managed appropriately to mitigate, in part, atmospheric increases in carbon dioxide which leads to global warming? How do climate parameters and land management issues influence rates of carbon sequestration in tropical wetlands?

Unfortunately, our current understanding of how environmental conditions and their fluctuations impact on carbon and water fluxes in tropical wetlands is limited, due largely to the paucity of measurement capacity.

However, there are an increasing number of measurements made on papyrus swamps, and what follows is a review of this limited information from two papyrus swamps in the Nile catchment and a complex wetland area in the Okavango Delta, Botswana.

The first measurements of carbon fluxes of a papyrus canopy were made on a lake-fringing papyrus swamp on the northwestern shore of Lake Naivasha in Kenya. The papyrus mat at the study site extended 500 m into the lake where it was floating, but it was grounded at the lake margin. During the period of measurements the lake was going through one of its natural cycles of draw-downs. The effect of this was to expose an increasing amount of the rhizome mat and associated detritus to the atmosphere, although the root systems remained continuously waterlogged. The fluxes of carbon were measured under water inundated and drawdown conditions and it was shown that under inundated conditions the papyrus swamp was a significant sink for carbon but when the rhizomes and detritus were partly exposed to the atmosphere during drawdown, there is a net ecosystem loss of CO_2 and the swamp became a carbon source (Jones and Humphries, 2002). During periods of inundation the papyrus wetland sequestered as much as 1.5 kg C m^{-2} y^{-1}, while during periods of hydrological drawdown the wetland released approximately 1.0 kg C m^{-2} y^{-1} (Figure 4.3).

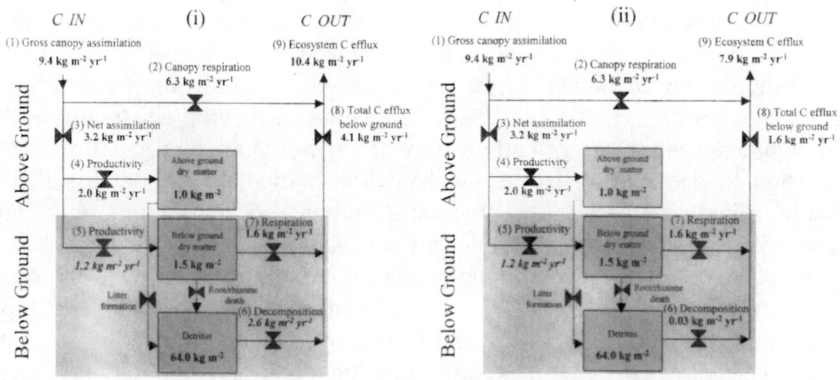

Exposed: Net C loss of 1.0 kg m^{-2} y^{-1} Inundated: Net C gain of 1.5 kg m^{-2} y^{-1}

Figure 4.3 Pools and fluxes of carbon for a papyrus ecosystem (i) when detritus and rhizomes are exposed to the atmosphere under conditions of hydrological draw-down and (ii) when detritus and rhizomes are inundates by water. Rectangles represent quantities and valve symbols represent flows. (1) Gross canopy photosynthesis is modelled. (2) Canopy respiration is measured. (3) Net assimilation is gross assimilation minus canopy respiration. (4) Above ground productivity is measured. (5) Below ground productivity is net assimilation minus above ground productivity. (6) Detritus decomposition is measure. (7) Below ground respiration is measured. (8) Total efflux is below ground respiration plus decomposition. (9) Ecosystem efflux is canopy respiration plus below ground efflux.

Source: Jones and Humphries (2002)

A second campaign of eddy covariance measurements were made on the Kirinya West wetland, a lagoon-based well-established papyrus wetland on the northern shore of Lake Victoria, Uganda. The wetland covers an area of 41,000 m^2 and is connected to Lake Victoria via the Napoleon Gulf (Saunders et al., 2007). Eddy covariance measurements were made over a period of six months and showed remarkably uniform patterns of diurnal carbon fluxes over the entire measurement period. Numerical integration of the net ecosystem exchange flux measurements showed that the papyrus ecosystem in this wetland sequestered approximately 0.48 kg C m^{-2} y^{-1} (Saunders et al., 2007). These relatively long term measurements on a stable papyrus ecosystem showed that papyrus wetlands have the potential to assimilate significant amounts of carbon under favourable conditions that allow for the development of long-term carbon stores in the detritus and peat deposits associated with these wetlands.

A third campaign of eddy covariance measurements have been used to assess the impacts of changes in management of papyrus wetlands on their carbon cycle (Saunders et al., 2012). In this study the impacts of agricultural encroachment on to pristine papyrus wetlands have been assessed. The pristine wetland was an area in the same wetland as assessed by Saunders et al. (2007) and the encroached area was used to cultivate cocoyams (*Colocasia esculenta*), an important staple crop grown for their edible underground stems. Population pressures in the area have resulted in the continued large scale removal of the endemic papyrus vegetation, mainly from the shallower margins of the papyrus wetland, in favour of food crops such as cocoyam. This form of land use change has been predicted to result in significant emissions of carbon dioxide as the papyrus vegetation is cut and left to decompose, while dissolved and particulate carbon losses are likely to increase through the physical disturbance of the rhizome mat and associated detrital deposits. While the cocoyam crop acts as a net sink for carbon dioxide during its growing season, there are losses through the harvest and export of the cocoyam corms from the wetland area. The flux measurements showed that while the nearby papyrus wetlands act as a carbon sink, when carbon exports from the cocoyam crop biomass was accounted for these wetlands become a carbon source. These measurements show that undisturbed papyrus wetlands have the potential to act as significant carbon sinks, but that when disturbed they can rapidly become carbon sources. However, there is clearly a need to provide more accurate information to link knowledge of the spatial extent of papyrus wetlands, their relative carbon sinks and how susceptible they may be to future changes in both climate and anthropogenic pressures related particularly to changes in land use. Furthermore, the long term sequestration of the measured net carbon gain in peat deposits is at present far from clear as under the flow through conditions, which exist below the floating papyrus mats, the peats can be swept away and possibly oxidised further down-stream to release CO_2 to the atmosphere. In addition, further carbon losses occur when dissolved organic carbon derived from the papyrus is carried away by the flow through (Opio et al., 2015).

The complex wetland area in the Okavango Delta in Botswana consists of permanently flooded, floating swamps dominated by papyrus and seasonal floodplains flooded for three to six months per year when the vegetation consists largely of the seasonal grasses, *Panicum repens* and *Oryza longistaminata*. The eddy covariance measurements (Helfter et al., 2021) were made over a three-year period (August 2017 to August 2020) when the hydrology showed significant inter-annual variation as a result of fluctuations in inflow of water from the Okavango River. In 2019 the river had the lowest input into the Okavango Delta on record which was 40% smaller than in 2018. The decline in water influx reduced the extent of the flooded area by almost 50% between 2018 and 2019 and this had particularly adverse effects on the vegetation of the seasonal floodplains, which suffered severe drought stress. As a result, during 2019, both the perennial papyrus swamp and the seasonal wetlands switched from net sinks for atmospheric CO_2 to net sources. In the case of the seasonal wetlands this was probably due to water stress of the vegetation, but in the case of the perennial wetland it was likely caused by the drawdown of water in the lagoon the swamp occupied and losses of carbon as the exposed rhizomes and roots decomposed. These results highlight the complexity of the processes that control the carbon balance of wetland ecosystems and they also illustrate the overarching control that hydrology has on inter-annual variability (Helfter et al., 2021).

Carbon Storage in Tropical Peatlands

In wetlands, carbon storage, or sequestration, occurs when peat, which consists of partly decomposed plant material, remains below the water table under anoxic conditions. When significant depths of peat have accumulated they are referred to as peatlands. There is no absolute consensus on what defines peat but Ribeiro et al. (2021) suggest that there are two important criteria; the first is that the accumulated material has an organic matter content of at least 65% and secondly that it has a minimum depth of 30–40 cm. Globally, peatlands cover a total area of between 185–423 million hectares (Mha), which accounts for between 1.2% and 2.8% of the earth's land area. There are two types of peatlands, ombrotrophic and minerotrophic, which are divided on the basis of the origin of the nutrient inputs into the system. In ombrotrophic peatlands the source of water is exclusively from precipitation while minerotrophic peatlands receive mineral nutrients, mainly as runoff from surrounding land or groundwater. The most extensive peatland vegetation cover in the tropics appears to be composed of dense canopies of trees but in other regions the vegetation cover consists largely of fast-growing herbaceous sedges and grasses. The woody material from trees appears to be more likely to form more substantial peat deposits than the fast-growing sedges and grasses as the dead branches and roots of trees have high carbon-to-nitrogen and lignin-to-nitrogen ratios that degrade slowly under anoxic conditions (Ribeiro et al., 2021). The tree covered peatlands in the tropics are therefore the main areas where deep peat deposits are found. However, the tropical wetlands with herbaceous vegetation can accumulate deep deposits rapidly under stable hydrological conditions (Hodgkins et al., 2018).

Peatlands in the tropics are located mainly in South America, Southeast Asia and Central Africa but Ribeiro et al. (2021) have pointed out that there is considerable variation in the published estimates of their occurrence and distribution. For example, until a few years ago southeast Asia was considered to have the most extensive peatland carbon reserves (Page et al., 2011), but recently, large intact peatlands have, for the first time, been described in South America and Africa. In Africa, most surprisingly, a previously unidentified complex in the Cuvette Central region of the Congo Basin has been identified by Dargie et al. (2017) and shown to be the largest intact tropical peatland to date. As a result of these discoveries estimates of tropical peatlands have been revised upwards to cover 90–120 Mha, which is 23–30% of the total area of peatlands globally. By combining estimates of peatland area with measurements of peat depth it has been estimated that the carbon storage in this region is approximately 29% of the total tropical carbon stock (Dargie et al., 2017) and about 5% of the estimated global peat carbon stocks. These massive increases in estimated area and amounts of peat deposits were initially treated with some caution as they were based on a small amount of field measurements made in very inaccessible areas and using satellite remote sensing that needed to be ground-truthed. Depth measurements are particularly difficult to make and require intensive field work and coring to determine depth. However, recent surveys of these peatlands have found that the peat can be up to 6 m deep and it has been accumulating for as much as 20,000 years (Young et al., 2023). These peatlands are isolated from the main river systems and are considered to be ombrotrophic-like peatlands where the peat formation most likely began with a significant increase in regional precipitation. The peat is consistently found under two common vegetation types consisting of hardwood swamp forest and a palm-dominated swamp forest and within this region, large scale shallow basins between the slow running adjacent rivers have filled with peat over several centuries and slowly increased in thickness, particularly away from the river margins.

Estimates of the area of peatlands in the Nile Basin are contained in the 'The Nile Basin Initiative corporate report' (NBI, 2019). The report also contains information on the carbon stocks, land-use threats and practical strategies for the sustainable use and management of these peatlands. The Nile Basin Initiative is an intergovernmental partnership of ten Nile Basin countries established in 1999 to coordinate among the basin states the sustainable management and development of the shared Nile Basin water and related resources. The report estimates that the total area of peatlands and other organic soils in the Nile Basin amounts to about 3 Mha and that this area contains a peat carbon stock of 4.4–10 Gt of carbon. The most important concentration of peatlands in the Nile Basin lies in the Sudd wetlands, where the estimated area of peatland is approximately 1.58 Mha, which represents about 50% of the total peatland area and 37% of the total carbon stock of the entire Nile Basin.

Wetlands and Outgassing of Carbon Dioxide from Inland Waters

Although the knowledge of the carbon cycle of papyrus wetlands derived from the eddy covariance measurements begins to build a better picture of how papyrus ecosystems function at the individual wetland scale it is still unclear as to what the contributions of these wetland ecosystems are at the regional and continental scale. In order to construct a full carbon budget of wetlands at the regional scale, in addition to the contribution of wetlands to the carbon sink as peat storage described above, there is a need to better understand the carbon flow from the wetlands to the major rivers such as the White Nile that deliver significant amounts of organic and inorganic carbon from wetlands and uplands to the sea (Cole et al., 2007). This process has been described as a riverine 'pipe' transporting carbon to the ocean. However, measurements show that not all of the carbon reaches the ocean, as roughly twice as much enters from wetlands and uplands as is exported to the sea and the remaining carbon is either lost by outgassing (evasion) to the atmosphere after aquatic respiration (aerobic and anaerobic), or retained as sediment storage and sequestered (Engle et al., 2008). The delivery of carbon through the riverine drainage network to the oceans and outgassing to the atmosphere is the end result of a number of transformations and losses in aquatic systems during transport, and it is only recently that it has become clearer how significant perennial wetlands are to these processes (Abril et al., 2014).

Most of the original research on assessing the tropical contributions to outgassing from riverine networks originated from monitoring of greenhouse fluxes in the wetlands and rivers of the Amazon basin (Richey et al., 2002) and until recently the contributions of African inland waters to CO_2 outgassing has been largely unknown. However, Borges et al. (2015) have reported globally significant greenhouse gas emissions, including CO_2, from the Congo basin, but as yet there have been no measurements reported for the Nile Basin. The original suggestion of Richey et al. (2002) was that the carbon source for the CO_2 outgassing in the Amazon probably originated from flooded forests and was subsequently respired and outgassed downstream. The recognition that the forest exports carbon, which decomposes in water and is released as CO_2, helped to reconcile a number of conflicting results from studies of the carbon cycle and carbon budget, both in the Amazon and on a global scale. In the case of the Amazon basin, Richey et al. (2002) suggested that the evasion of CO_2 is driven primarily by in-stream respiration of organic carbon, fixed originally on land and along river margins, attributing the source of CO_2 to carbon that has previously fixed upstream by land plant photosynthesis, then transferred to soils, and subsequently transported downstream in run off. However, it has subsequently become clear that, certainly in the case of the Amazon River, the CO_2 outgassing is largely the responsibility of the Amazonian wetlands. The work of Abril et al. (2014) showed that wetlands pump large amounts of CO_2 into river waters in the floodplains of the central Amazon and the dissolved CO_2 can be transported long distances downstream before being emitted.

The flux of carbon between the floating and emergent vegetation of wetlands and the river water occurs through two distinct pathways; either dissolved CO_2 from the wetland is derived from root and sediment respiration and exported from within the wetland or it is derived from dead and decaying vegetation. In the water and sediments, microbial heterotrophic respiration continuously converts organic carbon to CO_2. This whole process depends on a continuous and sufficient flow of water to carry the organic material and dissolved CO_2 away from beneath the vegetation. If this does not happen then the organic material is deposited as detritus which eventually accumulates as peat. The rate of accumulation of peat is therefore dependant on a fine balance between the rate of dead organic matter production and the ability of the moving water to carry it away.

In conclusion, disentangling the relative contributions of wetlands and upland carbon to CO_2 emissions from inland waters is essential for better understanding of the role of tropical inland waters in the global carbon cycle and the consequences for potential feedbacks on climate warming driven by increasing atmospheric CO_2 concentrations. Knowledge in this area has increased significantly in recent years and in particular we now have a better understanding of the processes involved. However, the contribution that the processes make to the local and regional carbon budgets in the tropics is still poorly quantified. Certainly, the sequestration of carbon as peat in wetlands has the net effect of reducing atmospheric CO_2 emissions and the rate of global warming but this is counteracted by the release of the greenhouse gas methane. The net ecosystem radiative (warming) balance is the result of the balance between greenhouse gas capture and release, neither of which is currently possible to determine at local or regional scales (Candry et al., 2023).

Methane Emissions

As well as being important global carbon sinks, storing approximately 30% of global soil organic carbon, wetlands are significant sources of tropospheric methane (CH_4). Globally, natural wetlands release about 20% of global methane emissions, a greenhouse gas with 32 times more powerful global warming potential than CO_2 over a 100 year timeframe (IPCC, 2022). Methane accumulation in the atmosphere makes a large contribution to the global radiative budget and currently it is responsible for about 0.5°C of the present global warming since the industrial revolution (IPCC, 2022). The net impact of wetland ecosystem on the Earth's radiant energy balance is the result of the balance between greenhouse gas capture in the form of CO_2 and release of methane. Determining the ecosystem radiation balance and the way that it may change in the future is critical in understanding the role of wetlands in future changes in global temperatures and climate (Mitsch et al., 2010).

Global concentrations of atmospheric methane have increased steadily since 2007, after nearly a decade of stability, but the causes for the renewed increase are not yet fully understood. Atmospheric methane is produced in three ways: (i) pyrogenically from biomass burning, particularly wildfires, (ii) thermogenically by the breakdown of organic matter buried deep in the Earth's crust at high pressures and temperatures, and (iii) biogenically by the microbial breakdown of organic matter in anaerobic environments. These anaerobic environments range from the stomachs of ruminant animals to wastewater and landfill systems, marine sediments, rice paddies and wetlands. The relative contributions of the three mechanisms vary with geography and in the case of methane from wetlands their largest contribution comes from the tropics where the three largest contributing regions for emissions are Brazil, Southeast Asia and Equatorial Africa, all located on the equator (Stavert et al., 2022). Emissions across tropical Africa are highly seasonal, with peaks in monthly emissions strongly correlated with the extent of the seasonally flooded grasslands. Lunt et al. (2019) found that around a third of the increase in global emissions has been due to a short-term increase in emissions between 2011 and 2015 from the Sudd wetlands. They used satellite observations of atmospheric methane concentrations and found that the higher methane emissions were most likely due to increased flow from the White Nile, which resulted in an increase in the extent of wetland flooding and consequentially higher methane production. It was suggested that the increase in flow of the White was due to higher rainfall over Lake Victoria, an increase in lake level and a substantial increase in outflow.

One of the main difficulties in determining wetland methane emissions at a regional scale is that their extent is difficult to quantify spatially and temporally and because of disagreement between different measures of wetland areas this equates to a wide range of emission estimates. Furthermore, as observational studies of methane emissions are usually made at point locations there are uncertainties around the contributions that differences in wetland vegetation composition make, particularly in accurately making measurements of emissions from either ebullition (gas bubble release), or plant-mediated transport and how they vary with vegetation composition and plant development (Were et al., 2021; Helfter et al., 2022; Vroom et al., 2022). The uncertainty in African methane emissions is also compounded by uncertainties associated with emissions from seasonal fires and agricultural methane emissions, particularly enteric fermentation from livestock. Despite the uncertainties, an ensemble of wetland emissions models has indicated that African wetland methane emissions represent approximately 12% of annual global wetland emissions. These emissions are concentrated in the sub-Saharan tropics and the most important sources are the Congo Basin located in Central Africa and the Sudd in South Sudan (Lunt et al. 2019).

The environment of wetland flooded detritus and soils exists as a dynamic equilibrium which is maintained by the availability of oxygen at the surface and buried organic carbon as a source of reducing power at increasing depth. There appears to be a particular sequence of reducing reactions as the environment becomes more reducing with depth. After oxygen is depleted by aerobic respiration near the surface, denitrification begins as the redox falls. The order of the subsequent anaerobic microbial processes is determined by the declining yield of metabolic energy from the reactions. Denitrifying bacteria use nitrate as an alternative electron acceptor during the oxidation of organic matter. When nitrate is depleted, the reduction of manganese and iron follows. Below the depth of iron reduction, the redox potential progressively drops to the point where sulphate reduction and methanogenesis begins. Sulphate reduction results in the escape of hydrogen sulphide from the soil but less than the rate of production because it reacts with other soil components including ferrous iron to precipitate ferrous sulphide to give the characteristic black colour to anaerobic soils. These reactions are all performed by specific anaerobic bacteria which reside in the detritus and soils. As a result, wetlands produce globally significant amounts of the two gases, hydrogen sulphide and methane, through organic matter decomposition (Candry et al., 2023).

Methane production in freshwater environments takes place by acetate splitting. The acetate compounds are produced from plant celluloses by fermenting bacteria that co-exist at the same depth as the methanogens. Methanogenesis is limited by the supply of labile organic matter and it has been demonstrated that the methane flux from some wetlands is directly correlated with the net ecosystem production. These measurements have shown that about 3% of net ecosystem production in the wetlands escapes to the atmosphere as methane gas. However, methane emissions from wetlands have also been shown to be influenced by vegetation characteristics including anatomy, phenology, species composition and plant density. Organic matter from different species may possess differing degrees of recalcitrance which influences the rate of decomposition. This suggests that CO_2 and CH_4 fluxes from different forms of vegetation can vary spatially.

The well-developed intercellular spaces forming aerenchyma in the roots and culms of papyrus provide a low resistance passage for methane to diffuse from the sediments and roots where it is formed to the atmosphere. This route is probably the most important for methane emissions from papyrus although significant amounts of methane are released from the sediment as gaseous bubbles. Were et al. (2021) compared the CH_4 fluxes from soils of papyrus-, *Typha*- and *Phragmites*-dominated wetlands in Uganda in different seasons of the year. They found that only *Phragmites* showed any seasonal differences, with lower methane emissions during the dry season. They concluded that water level was the main driver of the CO_2 and CH_4 fluxes from the wetlands, with low CO_2 and higher CH_4 fluxes during the high-water levels and the reverse at low-water levels.

Helfter et al. (2022) suggested that there is a need to consider emission mechanisms and processes to better understand the controls on methane emissions as fluxes of CH_4 to the atmosphere result from a complex of processes associated with the production, oxidation and transport that show temporal dynamics that are linked to vegetative processes and growth cycles. Production of methane is controlled by the availability of carbon for methanogenesis from either plant litter or photosynthate and this may contribute to the observed daily and seasonal patterns of emissions. However, the level of methane oxidation in the rhizosphere, which is modulated by oxygen diffusing through the plant aerenchyma, will also influence emissions and although plant mediated transport of CH_4 may be the dominant transport mechanism it may also be species dependent.

Three years of measurements of land-atmosphere exchange of CH_4 in permanent and seasonally flooded wetlands in the Okavango Delta, Botswana demonstrated that the net CH_4 emissions are controlled by seasonal changes in regional hydrology but the control was mainly imposed by the seasonal growth and development of the wetland vegetation (Helfter et al., 2021). The seasonal flooding regulated vegetation composition with papyrus domination of the permanent swamp and *Panicum repens*, the dominant grass in the seasonal swamp. They found that the largest CH_4 emissions were from the perennial swamp and that on a monthly basis the fluxes were linearly and positively correlated with maximum gross primary production as an indicator of the maximum photosynthetic potential of the papyrus stand. This is consistent with several other studies that suggested that methane production is stimulated by carbon deposition in the rhizosphere from photosynthesis. The data suggested that the largest CH_4 fluxes for an active papyrus stand occurs when the proportion of mature plant reaches a maximum. It also suggests that the dominant control of the seasonal cycle of CH_4 emissions is probably carbon deposition in the rhizosphere. However, there is also a daily cycle with emissions of CH_4 being 52% larger at night than during the day. This suggests that estimations of daily or longer emissions from daytime measurements would lead to significant underestimations. It is speculated that the reason for this could be reduced CH_4 oxidation at night due to reduced O_2 movement to the rhizosphere.

Scaled estimates of CH_4 emissions for the African continent, based on the flux measurements made in the Okavango Delta, indicate that the 0.1% land cover by papyrus accounts for 6% of the continent's total CH_4 emissions. However, using the annual minimum and maximum range these estimates could range from 2% to 10% of the continent's budget, which according to

Helfter et al. (2022) exemplifies the magnitude of the uncertainties on emissions from tropical vegetated wetland and the need to better constrain them.

Hydrological Cycle in Papyrus Wetlands

In the hydrological cycle, water moves through the stages of uptake by the plant, evaporation from the plant in the process of transpiration, condensation in the atmosphere, precipitation as rain and infiltration into the soil or runoff. The biogeochemical reactions in wetlands are linked to the reactions occurring in the surrounding terrestrial environment by the movements of surface waters, mainly in the form of rivers which carry runoff from upland ecosystems.

Water loss by evapotranspiration (the combination of evaporation from open water and transpiration from vegetation) from areas of emergent vegetation can have a significant impact on the hydrology of wetlands and associated water bodies. The productivity of papyrus wetlands is strongly coupled with the localised hydrological processes, where seasonal and inter-annual fluctuations in water level have a crucial effect on net primary productivity (Saunders et al., 2014). The flood pulsing concept has been increasingly recognised as highlighting the importance of the seasonal hydrological pulses reflecting seasonal patterns of rainfall (Rebelo et al., 2012). When considering the amount of evapotranspiration, the question as to whether the presence of aquatic vegetation leads to an increase or decrease in total evaporative losses from an area of water body has been a topic of debate for many decades, and there still appears to be no clear answer. There is certainly some evidence that the presence of papyrus under some circumstances does not accelerate evapotranspiration from wetlands.

Few measurements of evapotranspiration from papyrus swamps have been made and of those that have, they have they have produced somewhat contradictory information. Mohamed and Savenije (2014) pointed out that while some estimates had shown that swamp evapotranspiration in the Sudd was similar to open-water evaporation, Rijks (1969), who made measurements on a mature stand of papyrus, found the water loss was on average 40% less than evaporation than would occur from adjacent open water. There are also conflicting results from direct measurements using eddy covariance methodology, which was described above for measuring carbon dioxide fluxes, where the evapotranspiration from a papyrus swamp on Lake Naivasha, Kenya was found to be only 64% evaporation from open water (Jones and Humphries, 2002) but at the Kirinya wetland in Uganda, papyrus evapotranspiration was estimated to be 25% higher than evaporation from open water (Saunders et al., 2007). Unfortunately, the reasons for these apparent contradictions in the measurements of rates of evapotranspiration are not clear although we know that they are critical in controlling the hydrological cycle of the wetlands. Furthermore, they appear to have been selectively used to justify proposals to regulate the river access or water flowing through the wetland swamps of the Nile. The prime example being the plan to construct the Jonglei canal to bypass much of the Sudd swamp, which was predicated on the basis that this would ensure that more water was able to continue down-stream rather than be lost in evaporation.

Most researchers conclude that a major factor influencing the rate of evapotranspiration is the height of the vegetation so that canopies of tall emergent grass type leaves tend to increase evaporative loss whereas extensive canopies of short emergent vegetation tend to reduce evapotranspiration from such water bodies. However, these conclusions do not take account of any physiological differences between vegetation types, in that C_4 species have higher water use efficiencies than the C_3 species. It might be anticipated therefore that evaporative loss from canopies of C_4 vegetation, like papyrus, would be lower than water losses from C_3 vegetation of a similar stature, such as *Typha latifolia*.

Research on the hydrology of papyrus wetlands has been limited but, because of the contested proposal to build the Jonglei canal there were a series of studies aimed at assessing the likely effects of the canal on the Sudd region and reported in detail in *The Jonglei Canal: Impact and Opportunity*, edited by Paul Howell, Michael Lock and Stephen Cobb and published in 1988. Hydrological studies showed that, in the case of the Sudd wetland, the relatively stable precipitation and evaporation means that changes in the wetland hydrology are largely dominated by the upstream climate over the equatorial lakes, principally Lake Victoria (Mohamed & Savenije, 2014). The water balance of the Sudd wetland over the last century shows and increase as a direct result of the dramatic increase of the White Nile flow during the 1960s. High rainfall over Lake Victoria in 1960/61 increased the lake outflow substantially and this led to a doubling of the wetland area during the second half of the century. This substantial increase in wetland area implies more wetland evaporation which has implications for water resources conservation projects.

The loss of water vapour from vegetation is largely through stomatal pores on the leaf surface. The stomatal pore opening is controlled by a number of environmental and physiological processes and in the case of papyrus the vapour pressure deficit of the air appears to be particularly important. When the air is dry at high temperatures in the middle of the day the stomatal pores of papyrus have been shown to partially close (Jones and Muthuri, 1984). This reduces the rate of water loss in transpiration and may be an explanation for the control that papyrus vegetation has in regulating canopy evapotranspiration during the day.

Information on the long-term patterns of the wetland hydrology and in particular the wetland area is of vital importance for future plans of wetland management and the socio-economic development of the regions. The control of evapotranspiration by papyrus swamps is probably influential at the regional scale because of the size of some of the swamps. Ultimately, this will influence the regional moisture availability, cloud formation and radiation intensity. Evaporation leads to localised cooling due to the latent heat of vaporisation but if this is reduced by drainage the increase in sensible heat absorption by the ground results in higher air temperatures.

Mohamed and Savenije (2014) analysed the long-term trends of the Sudd hydrology from 1900 to 2000. They found no statistically significant change of precipitation and actual evaporation rates but there was a rise of the maximum temperature by 0.6°C in 100 years, and a larger increase if minimum temperature of 1.5°C for the same period. In addition, there was a decreasing trend of 10% during the second half of the century and a significant decreasing trend of measured sunshine by 10%. They concluded that the compensation of higher temperature by lower relative humidity, plus the lower solar radiation, resulted in no change of the actual evaporation for the wetland. Lower solar radiation is attributed to increased convective clouds because of the hotter wetland environment.

Mohamed and Savenije (2014) concluded that the wetland hydrology of the Sudd wetland is largely the result of the upstream climate over the Equatorial lakes which influences their outflow because there has been relatively little change in precipitation and evaporation over the Sudd wetland. Consequently, the significant increase in the water balance of the Sudd over the twentieth century is a direct result of the dramatic increase of the White Nile flow during the 1960s. High rainfall over Lake Victoria in 1961/1962 increased the lake outflow substantially, and subsequently doubled the wetland area during the second half of the last century compared to the first half.

Future Hydrology of Papyrus Wetlands

In the future, the papyrus wetlands face two major changes which will influence their hydrology. They are a changing climate alongside rapidly rising pressures on water resources due to population and economic growth. This will be further complicated by a range of politically complex transboundary water management issues. In recent years there has been increasing importance attached to analysing the impact of projected climate change on water resources with the prospect of proposing adaptation strategies. The analyses have a framework which can be summarised as follows (a) choosing scenarios from the IPCC (International Panel on Climate Change) reports on emission scenarios of greenhouse gas emissions, which depend on the future economy and energy use scenarios, (b) applying a combination of global circulation models (GCM), (c) downscaling the GCM outputs to the river basin scale, (d) using downscaled GCM outputs as inputs for hydrological models, and (e) analysis of the hydrological model results by comparison with current climate and hydrology. Potentially this allows the quantification of changes in floods and river flows as part of the future hydrological cycles (Di Baldassarre et al., 2011).

Climate models of future climate in the region indicate that by the end of the century the amount of rain falling in the Upper Nile Basin could increase by up to 20%. However, the projections also show that as well as increasing rainfall there will be more extreme events of heat and dry spells which will severely impact on the human populations of the region. It has been suggested that combined with increased populations this will cause chronic water scarcity for more than 40% of the population. Several modelling exercises suggest that the Nile Basin water supplies will be negatively impacted by the changing climate and this will, of course, be exacerbated by the increasing demand for water from an increasing population and the developing economies.

Intensity of rainfall and streamflow in the Upper Nile Basin have changed significantly in recent decades and are expected to continue to change as a result of changing climate. However, the choices of different scenarios, modelling and downscaling give variable results and therefore provided considerable uncertainty even in the direction of the changes. These inconsistencies in the climate projections mean that they become very unreliable and even misleading. As a consequence, their use for climate adaptation planning and river system management is very much open to question.

Interactions between Carbon and Water Fluxes

The assimilation of carbon in the carbon cycle of wetlands depends on the presence of water in sufficient amounts to maintain and optimise the physiological process of photosynthesis. The carbon and water cycles are interdependent as the release of water in transpiration is linked to the assimilation of carbon in photosynthesis. An example of a wetland system where carbon and water cycles of papyrus strongly influence each other is the Okavango Delta referred to above. The Okavango Delta is fed by the Okavango River and receives pulsed flooding of 8.5 ± 2.0 Gm^3 $year^{-1}$ through river discharge originating on the Angolan Highlands. During the rainy season (October to March) an additional 30% of the annual flow is received as rainfall. The floodwaters travel from the main inlet at Mohembo to the main outlet at Maun (250 km) over four months. Peak flood typically occurs in August, but the maximum annual extent depends on the combined inputs from the floodwater and precipitation which vary from year to year. The flooding regimes establish marked vegetation zones in the delta but the perennial swamps are dominated by papyrus. The recent monitoring of carbon dioxide and methane emissions using eddy covariance (Helfter et al., 2021) over three consecutive years has found that for the majority of the time the swamp was a carbon sink. However, there was very high variability driven by large difference in water influx. The changes in water influx were reflected in changes in extent of the total flooded area which shrunk by 50% for two of the three years. The changes in water supply showed that the papyrus ecosystem CO_2 fluxes are very dependent on the level of flooding with canopy photosynthesis declining rapidly as the dry conditions developed. In general, the complexities of the ecosystem fluxes of CO_2 are a challenge in interpreting the controllers of the fluxes although it clear that hydrology is the overarching control of inter-annual and long-term variability.

It is clear from these measurements that the fate of wetland vegetation in terms of distribution, abundance and species succession, and by extension the fate of the associated carbon budgets are directly impacted by the climate anomalies, in particular variable precipitation patterns and anthropogenic disturbances, such as water abstraction and wetland drainage. The combined effect is to alter the wetland hydrology ultimately in terms of the observed CO_2 flux measurements.

In the case of the Okavango Delta papyrus swamps the sensitivity of the carbon sources and sinks to drought and the large inter-annual variation in flooding combine to make a seasonally flooded papyrus ecosystem volatile in terms of their net ecosystem exchange of carbon dioxide and this is likely to intensify as precipitation patterns change in the future.

Conclusion

The productivity of papyrus wetlands is strongly coupled with the localised hydrological processes, where seasonal and inter-annual fluctuations in water level and availability have a crucial effect on net primary productivity. In order to understand the spatial and temporal variability of these processes we require a significant increase in the currently very limited number of direct measurements of the fluxes in the carbon and water cycles of papyrus wetlands. The need for this information has been highlighted by the fact that important ecosystem services provided by wetlands are carbon and water related. The accumulation of peat, preserves and accumulates organic carbon assimilated by the papyrus vegetation in the long term, but this depends on the maintenance of stable conditions below the papyrus rhizome mat. The conditions that ensure the long-term retention of the papyrus detritus have not been sufficiently understood to allow an assessment of the spatial variation in papyrus peat deposits. However, it has been estimated that under favourable conditions the papyrus wetlands have the potential to store more carbon than the equivalent area of forests (Saunders et al., 2007). Water related ecosystem services are the ability of wetlands to accumulate and store large amounts of water, which allows for stable release of water downstream and reduces the dangers from flash floods. Together, carbon related and water related ecosystem services are important contributors to the full range of ecosystem services provided by wetlands. A more detailed analysis of the contribution of papyrus wetlands to the supply of ecosystem services to human society will be considered in Chapter 6.

References

Abril, G., *et al.* (2014). Amazon River carbon dioxide outgassing fuelled by wetlands. *Nature*, 505, 395–398.

Borges, A.V., Darchambeau, F., Teodoru, C.R., Marwick, T.R., Tamooh, F., Geeraert, N., Omengo, F.O., Guérin, F., Lambert, T., Morana, C., Okuku, E., and Boullion, S. (2015). Globally significant greenhouse-gas emissions from African inland waters. *Nature Geoscience*, 8, 637–642.

Candry, P., Abrahamson, B., Stahl, D.A. and Winkler, M.K.H. (2023). Microbially mediated climate feedbacks from wetland ecosystems. *Global Change Biology*, 29, 5169–5183.

Cole, J.J., Prairie, Y.T., Caraco, N.F., McDowell, W.H., Tranvik, L.J., Striegl, R.G., Duarte, C.M., Kortelainen, P., Downing, J.A., Middelburg, J.J. and Melack, J. (2007). Plumbing the global carbon cycle: Integrating inland waters into the terrestrial carbon budget. *Ecosystems*, 10, 172–185.

Dargie, G.C., Lewis, S.L., Lawson, I.T., Mitchard, E.T., Page, S.E., Bocko, Y.E. and Ifo, S.A. (2017). Age, extent and carbon storage of the central Congo Basin peatland complex. *Nature*, 542, 86–90.

Di Baldassarre, G., Elshamy, M., van Griensven, A., Soliman, E., Kigobe, M., Ndomba, P., Mutemi, J., Mutua, F., Moges, S., Xuan, Y., Solomatine, D. and Uhlenbrook, S. (2011). Future hydrology and climate in the River Nile Basin: A review. *Hydrological Sciences Journal*, 56, 199–211.

Engle, D.L., Melack, J.M., Doyle, R.D. and Fisher, T.R. (2008). High rates of net primary production and turnover of floating grasses on the Amazon floodplain: Implications for aquatic respiration and regional CO_2 flux. *Global Change Biology*, 14, 369–381.

Helfter, C., Gondwe, M., Murray-Hudson, M., Makati, A. and Skiba, U. (2021). From sink to source: High inter-annual variability in the carbon budget of a Southern African wetland. *Philosophical Transactions of the Royal Society A*, 380, 20210148.

Helfter, C., Gondwe, M., Murray-Hudson, M., Makati, A., Lunt, M.F., Palmer, P.I. and Skiba, U. (2022). Phenology is the dominant control of methane emissions in a tropical non-forested wetland. *Nature Communications*, 13, 133.

Hodgkins, S. B., *et al.* (2018). Tropical peatland carbon storage linked to global latitudinal trends in peat recalcitrance. *Nature Communications*, 9, 3640.

Howell, P., Lock, M. and Cobbs, S. (eds). (1988). *The Jonglei Canal: Impact and opportunity*. Cambridge: Cambridge University Press.

IPCC. (2022). *Climate change 2022: Impacts, adaptation, and vulnerability*. Contribution of Working Group II to the Sixth Assessment Report of the Intergovernmental Panel on Climate Change. Cambridge: Cambridge University Press.

Jones, M.B. (1986). Wetlands. In N.R. Baker & S.P. Long (Eds). *Photosynthesis in contrasting environments*. Amsterdam, Elsevier.

Jones, M.B. (1987). The photosynthetic characteristics of papyrus in a tropical swamp. *Oecologia*, 71, 355–359.

Jones, M.B. and Muthuri, F.M. (1984). The diurnal course of plant water potential, stomatal conductance and transpiration in a papyrus (*Cyperus papyrus* L.) canopy. *Oecologia*, 63, 252–255.

Jones, M.B. and Muthuri, F.M. (1985). The canopy structure and microclimate of papyrus (*Cyperus papyrus*) swamps. *Journal of Ecology*, 73, 481–491.

Jones, M.B. and Muthuri, F.M. (1997). Standing biomass and carbon distribution in a papyrus (*Cyperus papyrus* L.) swamp on Lake Naivasha, Kenya. *Journal of Tropical Ecology*, 13, 347–356.

Jones, M.B. and Humphries, S.W. (2002). Impacts of the C_4 sedge *Cyperus papyrus* L. on carbon and water fluxes in an African wetland. *Hydrobiologia*, 488, 107–113.

Lunt, M.F., Palmer, P.I., Feng, L., Taylor, C.M., Boesch, H. and Parker, R. J. (2019). An increase in methane emissions from tropical Africa between 2010 and 2016 inferred from satellite data. *Atmospheric Chemistry and Physics*, 19, 14721–14740.

Mitsch, W.J., Nahlik, A., Wolski, P., Bernal, B., Zhang, L. and Ramberg, L. (2010). Tropical wetlands: Seasonal hydrologic pulsing, carbon sequestration, and methane emissions. *Wetlands Ecology and Management*, 18, 573–586.

Mohamed, Y. and Savenije, H.H.G. (2014). Impact of climate variability on the hydrology of the Sudd wetland: Signals derived from long term (1900–2000) water balance computations. *Wetlands Ecology and Management*, 22, 191–198.

Muthuri, F.M. and Kinyamario, J.I. (1989). Nutritive value of papyrus (*Cyperus papyrus*, Cyperaceae), a tropical emergent macrophyte. *Economic Botany*, 43, 23–30.

NBI. (2019). *Nile Basin Initiative corporate report 2019*. https://nelsap.nilebasin.org/index.php/en/media-items/documents-publications/28-nbi-corporate-report-2019-1/file.

Opio, A., Jones, M.B., Kansiime, F. and Otiti, O. (2015). Dissolved organic carbon in a tropical wetland dominated by *Cyperus papyrus*. *Wetlands Ecology and Management*, 23, 1033–1038. doi:10.1007/s11273-015-9437-z.

Page, S.E., Rieley, J.O. and Banks, C.J. (2011). Global and regional importance of the tropical peatland carbon pool. *Global Change Biology*, 17, 798–818.

Rebelo, L.-M., Senay, G.B. and McCartney, M.P. (2012) Flood pulsing in the Sudd wetland: Analysis of seasonal variations in inundation and evaporation in South Sudan. *Earth Interactions*, 16, 1–19.

Ribeiro, K., Pacheco, F.S., Ferreira, J.W., de Sousa-Neto, E.R., Hastie, A., Krieger Filho, G.C., Alvalá, P.C., Forti, M.C. and Ometto, J.P. (2021). Tropical peatlands and their contribution to the global carbon cycle and climate change. *Global Change Biology*, 27(3), 489–505.

Richey, J.E., Melack, J.M., Aufdenkampe, A.K., Ballester, V.M., and Hess, L.L. (2002). Outgassing from Amazonian rivers and wetlands as a large tropical source of atmospheric CO_2. *Nature*, 416, 617–620.

Rijks, D.A. (1969). Evaporation from a papyrus swamp. *Quarterly Journal of the Royal Meteorological Society*, 95, 643–649.

Saunders, M.J., Jones, M.B. and Kansiime, F. (2007) Carbon and water cycles in tropical papyrus wetlands. *Wetlands Ecology and Management*, 15, 489–498.

Saunders, M.J., Kansiime, F. and Jones, M.B. (2012) Agricultural encroachment: Implications for carbon sequestration in African tropical wetlands. *Global Change Biology*, 18, 1312–1321.

Saunders, M.J., Kansiime, F. and Jones, M.B. (2014). Reviewing the carbon cycle dynamics and carbon sequestration potential of *Cyperus papyrus L.* wetlands in tropical Africa. *Wetlands Ecology and Management*, 22, 143–155.

Schlesinger, W.H. (1997) *Biogeochemistry: An analysis of global change*. San Diego, CA: Academic Press.

Stavert, A.R., *et al.* (2022). Regional trends and drivers of the global methane budget. *Global Change Biology*, 28(1), 182–200.

Vroom, R.J.E., Van Den Berg, M., Pangala, S.R., Van Der Scheer, O.E. and Sorrell, B. K. (2022). Physiological processes affecting methane transport by wetland vegetation: A review. *Aquatic Botany*, 182, 103547.

Were, D., Kansiime, F., Fetahi, T., Cooper, A. and Jjuuko, C. (2019). Carbon sequestration by wetlands: A critical review of enhancement measures for climate change mitigation. *Earth Systems and Environment*, 3, 327–340.

Were, D., Kansiime, F., Fetahi, T. and Hein, D. (2021). Carbon dioxide and methane fluxes from various vegetation communities of a tropical natural freshwater wetland in different seasons. *Environmental Processes*, 8, 553–571. doi:10.1007/s40710-021-00497-0.

Young, D.M., Baird, A.J., Morris, P.J., Dargie, G.C., Mampouya Wenina, Y.E., Mbemba, M., Boom, A., Cook, P., Betts, R., Burke, E. and Bocko, Y.E. (2023). Simulating carbon accumulation and loss in the central Congo peatlands. *Global Change Biology*, 29, 6812–6827.

5 Threats to Papyrus Wetlands

Anthropogenic Impacts, Altered Hydrology and Climate Change

Introduction

This chapter describes the impacts of the ever-increasing human population and rapidly changing climate on papyrus-dominated wetlands with a particular focus on the Nile catchment. Human population growth from the arrival of the earliest settlements of cattle herders in the Nile Valley, around 7,000 years ago, to the present has posed increasing threats to papyrus wetlands (Zhao et al., 2022). We review how humans in their interactions with wetlands have impacted on papyrus vegetation and what the short and long term consequences have been for the papyrus-dominated wetlands. It is well established that the hydrological regime is the primary factor controlling the distribution of papyrus and its relative importance in different wetlands. Humans over the centuries have altered, and will continue to modify, the hydrology of the Nile Basin and this has had a major impact on the papyrus swamps in terms of their distribution and functioning. An additional threat to wetlands in more recent times is human induced climate change, which is predicted to have its most significant impacts through changes in the hydrology of the Nile Basin.

Many communities, along the course of the Nile, but particularly in the Lake Victoria Basin, rely entirely on waters extracted from the wetlands for drinking, cooking and washing. Others rely on material gathered in wetlands such as clay, sand and vegetation, in particular papyrus. However, the main impact of humans has been to clear the natural vegetation and replace it with agricultural crops with negative ecological consequences. Suitable, sustainable management of these wetlands is therefore essential for maintaining the livelihood of many of these communities adjacent to the wetlands. Wetlands are typically exploited by stake-holders with different interests who can lay claim to the wetlands water and associated land. Stake-holders include users who directly harvest the wetland's goods in a sustainable way, agricultural producers that drain and convert the wetlands to agricultural land, users that benefit from indirect wetland services such as storm abatement and flood mitigation, and individuals and amenity groups whose objective is to conserve the natural environment. In many cases it is

DOI: 10.4324/9781003131540-5

likely that the different interests of the stake-holders conflict, so that development of policy for the use of wetlands involves a complex series of trade-offs (Schuyt, 2005).

Wetlands have long been identified as one of the most threatened ecosystems globally, as a result of their drainage and conversion to agricultural land (Junk et al., 2013). Drained wetlands can provide nutrient rich soils with a high moisture content, often enabling farmers to produce crops all year round. In sub-Sahara Africa, the use of wetlands for agricultural production has been increasingly seen as a potential solution to food security challenges as other wetland services have been commonly perceived to have minimal value in comparison to the production of valuable food crops.

When considering the history of human impacts on wetlands of the Nile Basin, there is a clear temporal separation in the severity of impacts between the swamps of the Nile Delta and Lower Nile and those of the Upper Nile that form the headwaters of the White Nile. Historical evidence shows that the papyrus swamps of the Lower Nile Valley were subjected to heavy human impacts beginning as long as 6,500 years ago, while the admittedly much more fragmentary evidence that we have available on the Upper Nile suggests that the papyrus swamps were free of human threats and remained largely pristine until the middle of the last century. Human impacts on the Lower Nile and the Upper Nile Basin wetlands will be considered independently.

Impacts of Agricultural Development in the Lower Nile Region

The Nile Delta wetlands developed as the river's flood-waters flowed seasonally across the flat areas of adjacent land and into the Mediterranean Sea. The flood had seasonal pulses of flow which reflect the wet and dry periods from the river sources and in particular of the Blue Nile flowing from Ethiopia. The flood-waters away from the main channels flowed slowly enough to deposit silt and allow the development of vast areas of fast growing vegetation that is adapted to growing either rooted in the sediment or floating on the surface of deeper waters. Papyrus is supremely well adapted to these conditions as following germination of seeds deposited with the sediment it forms rooted seedlings on the sediment initially, but as the water levels rise during a flood pulse the small roots detach from the sediment and develop into larger floating rhizomes. It is from the floating rhizomes that the mature culms eventually develop forming vast areas of papyrus swamp extending over the open water (Terer et al., 2014).

The 'flood pulse concept' developed by Junk et al. (1989) describes the importance of the hydrological pulse to the functioning of wetlands in the region as the flood-waters deposits their burden of nutrient-rich silt across the floodplain. It proposes that the annual inundation pattern is primarily responsible for the high productivity of the biotic systems within the wetlands. The flood pulses that regulate the annual cycles in tropical wetlands were traditionally exploited by the early settlers to allow agriculture and annual grazing during the low water periods, and fisheries throughout the year. The high primary production makes wetlands important for what is known as 'flood recession agriculture' as well as for grazing of livestock. Recessional agriculture is defined as a form of agricultural cultivation that takes place on a floodplain that allows cultivators to successively grow crops in the formerly flooded areas after the water recedes.

This cycle of cultivation was probably used by Neolithic farmers in the Nile Delta from the time when a regular flood regime was established from the middle of the Holocene (around 3000 BCE) onwards. In the Nile Delta the subsequent evolution of irrigation cultivation was supported by a number of ingenious ways, particularly raising the water levels by extending erosional gullies to divert floodwater towards the low-lying marshy swamps along the outer margins of the flood plains. Later, the well organised Egyptian ruling elite were able to use an abundant labour force to build and maintain a sophisticated canal system to irrigate areas that could grow crops for much of the year.

The importance of riverine agriculture to the early Egyptians was that it offered great potential to dramatically increase agricultural yields. However, the cultivators had to cope with a largely unpredictable annual fluctuation in the amount of water that arrived in the flood pulses, which meant that there could be enormous variability in agricultural productivity from year to year. As one of the ancient world's great 'hydraulic civilisations' Egypt's prosperity was overwhelmingly tied to the annual cycle of the Nile's summer flooding and the failures of these floods were associated with major human impacts through its many millennia of recorded history. The summer floods, driven mainly by the monsoon rainfall in the Ethiopian highlands, began with rising waters observed at Aswan as early as June, peaking from August to September, and largely receding by the end of October, when the crop sowing began. Failures of the monsoon rains could have dramatic consequences which were sufficient to trigger economic and political instability and, for instance, trigger revolts against the Ptolemaic rule of Egypt (305–30 BCE) as documented by Manning et al. (2017). Dealing with year to year fluctuations, and in particular attempting to insure against low crop yields during the dry years, depended largely on increasing the area of cultivatable land by clearing papyrus swamps and if possible selecting for varieties of crops that appeared to be more tolerant of drought. It also increasingly depended on developing tools to aid in the preparation of ground before sowing the crops and improving the efficiency of harvesting and storage of the crops. This may have stretched to allowing for the storage of crop harvests for two years or more that could continue to supply food following crop failures. The redistribution of stores by boat also allowed for the relief of local shortages. Despite these efforts it appears that on longer time scales of a few decades up to centuries there were extended periods of severe drought associated with regional climate change, probably related to the displacement of the Intertropical Convergent Zone.

When the cattle herders arrived in the Lower Nile Valley and began to settle they would have had close contact with the delta wetlands as they lived mainly on the fertile floodplain and delta of the river. Close proximity to the river meant that the Nile Valley dwellers began to understand how the river swelled and diminished through its annual cycle and also how the channels and islands of the delta behaved over the generations. Initially, the herders most likely avoided the papyrus dominated wetlands and accessed water, particularly for their cattle, from the open faster flowing river. Subsequently, they appear to have abandoned the herding and changed to a more settled lifestyle, cultivating cereals. From then on they probably began to utilise the plentiful papyrus in the swamps and their settled lifestyle made it more likely that they would begin to manage the swamps sustainably, in order to extract the maximum benefits from them. The first uses would most likely have been to make simple boats for local transport and for some early forms of housing structures. However they later started clearing papyrus for crop production although they had none of the tools and draining technology of later years. For many years small areas of swamp were probably drained and cleared locally and the remaining swamps were harvested for local materials for boats, houses, matting and craft purposes. In time, in order to obtain the highest quality papyrus raw material for the manufacture of papyri, it was probably necessary to manage the swamps to ensure regular harvests of the freshly growing papyrus culms and to maintain easy access to the swamp for the harvesting of the culms (Said, 1993). Alongside the production of cereals, papyrus was growing in the permanently flooded, more stable backwaters, of the delta. The area of wetland available for papyrus cultivation during the peak in export of papyrus rolls from before the Ptolemy's until late into the Roman Empire, has been estimate by John Gaudet (2014), using a number of published sources, to be about 6,500 km^2. This was the area sufficient to supply, what was at that time, the global market for papyrus paper. What subsequently led to the rapid collapse in the production of papyrus paper was the wide availability of cheaper rag paper. Certainly, if papyrus paper was no longer required and there was no market for it, the papyrus swamps would have been cleared by the local community to create space for cultivation of more useful vegetation, particularly cereals, which were increasingly required for the expanding population and for the large export market. This was the beginning of a rapid decline in the use of any remaining papyrus swamps and we know from contemporary records that by the beginning of the nineteenth century papyrus was no longer to be found growing in significant amounts in the Lower Nile and Delta region.

Impacts of Altered Hydrology in the Lower Nile Region

On a geological time scale the River Nile as we know it today is a very recent development (Said, 1993). It is the result of the interconnections of several independent basins by rivers which developed during the last wet period which affected Africa after the retreat of the ice of the last glacial age 10,000 years ago. The Nile River dynamics have also shown both short (centennial) and longer (millennial) term hydroclimatic changes that have resulted in repeated episodes of river channel and floodplain contraction resulting from flow reduction (Foucault and Stanley, 1989). These repeated and extended periods of water stress must have had a profound and lasting impact on riverine societies and in particular floodwater farmers in the Nile Valley. To cope with these reductions in the reliability of the annual floods, the farmers were forced to develop increasingly sophisticated methods to lift water to nearby fields and maintain the supply of cereals on which the communities were dependent.

Up until the early part of the last century the changes in river hydrology were largely the consequence of climatic variations and particularly the changed patterns of rainfall both on an annual and millennial time scale. As we have seen, the cultivated reaches of the Lower Nile during the later Egyptian dynasties were managed to intensify agricultural production through controlled irrigation and drainage of the land. This was the first time that humans had a direct impact on the hydrology of the river, but since then the hydrology of the Nile, mainly along its lower reaches in Sudan and Egypt, has been variably modified by humans by the construction of dams, and where these modifications have occurred they have affected the ecology of the river and the land areas adjacent to it. The expected long-term benefits of constructing dams include flood management, water storage for irrigation and electric power generation, but these are partly counterbalanced by a number of undesirable side-effects. These include the deposition of suspended silts in the reservoir and the resultant reduction of nutrient rich silt deposits in the Nile Delta region, the need for additional fertilisers to compensate for the loss of silt, salt accumulation in soils downstream of the reservoir and an increase in coastal erosion along the delta. In addition there are the social consequence of displacement and resettlement of people whose homes and land has been submerged beneath the rising waters of the reservoirs (Williams, 2019).

The construction of dams along the Nile have always had the potential to cause conflict between the Nile Basin countries as a result of the impact the dams have on their water supplies and the need to allow a fair share of the water resources. The latest and potentially most intractable issues have been generated over the construction of The Grand Ethiopian Renaissance Dam (GERD) (see Chapter 3) which has been a source of tension between Egypt, Sudan and Ethiopia. Ethiopia, whose highlands provide more than 85% of the water that flows into the Nile and has long argued that it has the right under international law to manage resources within its own borders for its national development. For Egypt, on the other hand, there is an expected reduction in hydropower generation and reduced irrigation waters available when multi-year droughts occur (Basheer et al., 2023). The immediate issue is that water flow to the downstream countries might be negatively affected during the filling of the dam's reservoir because the storage capacity is 1.5 times the historical mean annual river flow. Extended negotiations over two decades between Ethiopia, Sudan and Egypt failed to produce an agreement and Ethiopia has blamed Egypt for the stalemate.

A critically important requirement is for the basin states to create an equitable water allocation policy that not only addresses 'historical rights', such as Egypt's long-held claims to two-thirds of the Nile flow, but also to the needs of upstream countries like South Sudan and Ethiopia, which require water to grow their economies. In the future, countries in the Nile Basin will need to work together to ensure the best use of the region's resources. There will be a need to trap and store rainfall in the wet, flood risk years, and equitably distribute the water in the dry years. Failure to work together to prepare could have catastrophic consequences.

Some progress on resolving these issues was made in October 2024 when the 'Nile Basin Cooperative Framework' – also known as the Cooperative Framework Agreement (CFA) came into force. The CFA was the result of over a decade of negotiations among all nine Nile riparian countries (Burundi, DR Congo, Egypt, Ethiopia, Kenya, Rwanda, Sudan, Tanzania and Uganda). The CFA presents a vision for how these states can come together to ensure the sustainable and equitable use of the Niles's resources. The key feature of the CFA is its promotion of transboundary cooperation and ensuring that all riparian states have a voice in the management of the river's resources. It establishes a mechanism for peaceful dispute resolution and aims to rectify historical imbalances in access to the Nile's waters and ensures that all Nile Basin countries can benefit from the shared resources. The CFA is seen as paving the way for the establishment of a Nile River Basin Commission (NRBC) which will have the primary objective of promoting and facilitating the implementation of the principles, rights and objectives provided for in the CFA and ensuring closer cooperation among the States and peoples of the Nile Basin to resolve disputes related to the river's water usage in the social, economic and cultural fields.

Anthropogenic Impacts on Wetlands of the Upper Nile Basin

While extensive papyrus-dominated wetlands had largely disappeared from the Lower Nile region from the Middle Ages onwards, they remained continuously present, and until the last half century largely undisturbed, from South Sudan and southwards to the upper reaches of the Nile Basin that form the headwaters of the White Nile. In fact, the continued presence of extensive papyrus swamps in the Sudd wetlands of South Sudan became very important in inhibiting the exploration of the upper regions of the Nile in the eighteenth and nineteenth centuries. Throughout this region the recent decline in wetlands area has been linked to the demands of the growing populations and their economic development. The population of the Lake Victoria Basin grew from an estimated 4.6 million in 1932 to 27.7 million in

1995 (Verschuren et al., 2002) although between 1900 and about 1930 it had not increased appreciably beyond the pre-colonial levels because of high natural mortality under early colonial rule, exacerbated by the impact of exotic diseases. Completion of the Uganda railroad by 1930 opened up the Lake Victoria region to settlement by expanding plantation agriculture for the export of commercial crops. Rapid population growth through immigration and improved health conditions started a pattern of large-scale deforestation and agricultural conversion of wetlands that continues to the present day when the Lake Victoria basin has one of the world's densest rural populations (Ngodhe et al., 2016). The population growth has also fuelled rapid urbanisation, industrial development and rural settlement. It is likely that in the past the wetlands, particularly those around the perimeter of Lake Victoria, were used by communities as a hedge against food shortages and did not involve wholesale conversion of wetlands. However, more recently the use of wetlands for agricultural production has been increasingly considered as a potential solution to food security challenges of the region and the complete drainage of wetland has become more common while other wetlands continue to be degraded. Agricultural demands are not only determined by population size but by also by increasing average income which together have increased demand for food and it is now estimated that converted wetlands provide 10–40% of the annual food needs of the rural population (Schuyt, 2005). In some circumstances the increasing demand for food and other material came at a time when droughts reduced food production from upland agriculture and intensified the demand for productive land. Wetlands were then seen as large, cheap and unencumbered land which can be easily developed (Willbroad et al., 2019). Consequently, wetlands have been increasingly reclaimed for intensive agricultural production, and in some cases for large scale farming through corporate or government activity (Ondiek et al., 2020). Further threats to the wetland areas around Lake Victoria relate to the production of palm oil where international funding has come from the World Bank and the International Fund for Agricultural Development. The possibility of selling their land increased the value of the formally low value farming plots to land users (Bikangaga et al. 2007). This encroachment has been greatest close to areas where roads provide easy access to markets but in more remote areas rapid population growth has led to increasing demand and acute shortages of land. The increasing demand has forced farmers to cultivate increasingly marginal land and often the only option is to drain papyrus swamps to grow food crops (Maclean et al., 2014). In other situations, where drainage is not possible this clearance has been to develop fish farming and rice farming. In some cases the cleared swamp area has been flooded by damming to control water levels for rice cultivation; for example the Yala swamp in Kenya.

In broad terms, the high productivity of wetlands now supports large numbers of poor people dependent on their ecosystem services for at least part of their livelihood. In Uganda it is estimated that over 50% of the total wetlands are under some form of human use. Governments of countries in the Lake Victoria basin have implemented a number of policies to support the concept of the wise use of wetlands in accordance with the Ramsar Convention that defines wise use as 'the maintenance of their ecological character, achieved through the implementation of ecosystem approaches within the context of sustainable development' (Ramsar Convention Secretariat, 2007). This recognises that some economic activities are inevitable but must be carried out in a sustainable way, and while sustainable ecosystem management is a long-term goal, survival is more immediate to poor people depending on ecosystem services for their livelihood (Mitchell, 2013). An example of how conservation efforts have provoked conflicts between land owners and policy makers comes from a review of perceived benefits of wetlands in the Nabugabo wetlands located in Central Uganda (Bikangaga et al., 2007). After receiving Ramsar status, conflicts arose in particular over the demarcation of areas for conservation and although conservation activities were viewed positively by some of the local community many others were sceptical as to whether it would help them as they remained convinced that the developments would severely limit their access to basic resources obtained from the wetlands.

Although the general perception has been that the areas of papyrus wetlands have declined significantly, particularly since the 1980s, there have been very few direct measurements which have monitored the loss over a number of years (van Dam et al., 2014). Those that have been made have generally shown that the rate of loss can be quite variable from one location to another but can range from 0.5 to 5% per year. As a result of this loss the areas of papyrus wetlands in some locations may have halved over a period of 25 years (van Dam et al. 2014). The direct causes of the habitat loss may be different from one location to another. For example, a survey of wetland changes in three different agri-ecological zones in Uganda carried out by Turyahabwe et al. (2013) found that very different factors led to wetland losses. In one area (Kyoga plains in central Uganda) there was increased growing of paddy rice, in a second (southwestern farmlands) grasslands for livestock grazing were expanded, and in a third (close to urban centres in the Lake Victoria Crescent) there was an expansion of vegetable growing to supply the local markets.

While the losses of papyrus wetlands have a number of socio-economic consequences, they also have very significant impacts on wildlife conservation. This is because the papyrus swamps host several threatened wildlife species such as the sitatunga antelope (*Tragalaphus spekei*) and African python (*Python seboe*), and a suite of papyrus specialist birds including the globally threatened papyrus yellow warbler (*Chloropeta gracilirostris*) and papyrus gonolek (*Laniarius mufumbiri*) (Maclean et al., 2014). The bird species are considered to be among the most threatened in east and central Africa. Using satellite imagery Owino and Ryan (2007) quantified drainage of papyrus swamp habitats in the wetlands of western Kenya and made assessments of the impact on eight species most closely associated with the wetlands. They found that between the periods of 1984–1987 and 1999–2001 the overall aerial extent of the papyrus wetland declined by 6.7% but the papyrus specialist birds underwent much greater declines. This was due in part to fragmentation effects and in part to geographical overlap between areas of highest population densities and areas that had experienced greatest habitat loss. It is suggested that the major conservation measure most likely to reduce the degree of threat facing papyrus-dwelling passerines would be to identify priority sites for conservation as it is neither practical nor appropriate to allocate resources to protect all the swamps and a high proportion of the threatened species populations are in fact hosted by a relatively small proportion of the swamp (Maclean et al., 2014). Clearly, this approach requires policies to be implemented at a local level and there are few examples of these governance structures actually in place.

The decline in areas of papyrus wetlands has been associated with dramatic ecological changes, which in the case of Lake Victoria has been particularly severe (Verschuren et al., 2002; Owino and Ryan, 2007; Isunju and Kemp, 2016). These changes included, in open areas of the lake, a massive increase in phytoplankton primary production, replacement of diatoms by cyanobacteria as the dominant group of planktonic algae and the eradication of several hundred species of endemic cichlid fishes after the 1980s population explosion of an introduced piscivore, the Nile perch (Schuyt, 2005; Odada et al., 2004). The observed changes in the phytoplankton community appears to be a consequence of the excess nutrient loading of the lake that results from deforestation and greatly intensified agriculture throughout the drainage basin and particularly around the margins of the lake and associated wetlands. Increased nutrient loading is not only the result of increased inputs but is a consequence of the absence of nutrient interception and sequestration by the papyrus swamp ecosystem that has been removed or severely degraded. Lack of long-term monitoring data as well as the complexity of the ongoing ecosystem changes have hampered the development of a clear understanding of the impact of disturbances on the current status of the wetlands. However, in the absence of monitoring data, the analysis of lake deposits in cores of offshore sediments has offered a powerful tool enabling the reconstruction of the succession and timing of the changes and the description of the environmental history of Lake Victoria over the last two centuries (Verschuren et al., 2002). These analyses show that the historical changes in phytoplankton productivity have been largely caused by excess nutrient loading and that the productivity increase matches human population growth and associated agricultural activity of the Lake Victoria basin. The conclusion from the analyses confirmed that with the current rate of growth of the local human population further degradation of the Lake Victoria ecosystem seems inevitable. Consequently, land-management strategies that severely restrict nutrient inputs to the lake and its tributaries need to be implemented on a basin-wide scale. Furthermore, the evidence is that the long-term sustainable ecology of Lake Victoria and other Upper Nile Basin wetlands and their ability to support their local populations are highly dependent on the continued presence of healthy and productive papyrus vegetation.

When undisturbed by human activity papyrus forms a fringing swamp of variable width around the perimeter of Lake Victoria which is a buffer between the land activities and the lake. The fringing swamp acts as a giant filter to remove effluents carried from the land, protecting the lake from eutrophication (Kansiime et al., 2007). The growing papyrus takes up nutrients from both the water and sediments and thus trap nutrients especially nitrogen and phosphorus. Nitrogen and phosphorus removed from the lake water includes temporary storage in growing culms and rhizomes and adsorption to the silt and detritus and longer term storage in deeper peat layers. Permanent removal of the nitrogen from the wetland ecosystem is through the process of denitrification in the anaerobic water surrounding the rhizomes.

The lake fringing papyrus swamps are largely floating and are subjected to fluctuating levels of Lake Victoria. Lake level fluctuations are the result of the lake's water balance, which consists of precipitation on the lake surface (~70%) and inflow from the tributary rivers (~20–30%) as input terms and outflow from the Nalubaale dam complex near Jinja (~20–30%) and evaporation from the lake surface (~70–80%). Lake precipitation and inflow control seasonal and interannual lake level variability as evaporation and outflow are characterised by lower variability. Outflow at the dam complex from the lake is managed as a function of what is known as the 'agreed curve', maintained under a treaty with Egypt to ensure that the outflow for Lake Victoria remains consistent with the natural state (Pietroiusti et al., 2024). Precipitation over the lake is considered to be the main cause of seasonal and interannual lake level fluctuations and seasonal variations are generally small compared to interannual variations.

The fringing papyrus swamps are most vulnerable to falls rather than rises in lake levels, referred to as drawdowns, as this leaves the papyrus rhizome mat stranded whereas the rhizome mat can float up as water levels rise. As an example, a rapid decline in levels of Lake Victoria in 2002–2006, resulted in a severe disruption of the fringing papyrus wetlands. The stranding cut off the papyrus from its water supply and much of it suffered severe water stress which reduced its growth. Kiwango and Wolanski (2008) were able to review one of the impacts resulting from this event. Papyrus wetlands have traditionally been essential locations for part of the life cycle of the most important fish, primarily Tilapia, caught by local wetland fishermen in Lake Victoria. Here, the water below the roots and rhizomes provides a refuge for the juvenile Tilapia fish. Asa result of the fall in water level the refuge of papyrus rhizomes was unavailable to the juveniles and as a result the population of Tilapia fish in the lake collapsed by as much as an estimated 80%. The Tilapia are the most important fish supplying the indigenous population whereas commercial fishing on Lake Victoria relies mainly on Nile Perch. Nile Perch appears to be less vulnerable than Tilapia to the effects of the drawdown and as a result the artisanal fisheries were more negatively impacted by lake drawdowns than the commercial fisheries.

A recovery from the 2000–2006 drawdown followed unseasonably large rainfall which occurred from November 2006 through to February 2007 and the lake level rose by 1 m in the first six months of 2007 and inundated the papyrus rhizome mats again. Much of the papyrus had survived 18 months of exposure by drawing on the water retained in the detritus peat and sediment below the rhizome mat. Kiwango and Wolanski (2008) suggested that if the lake water level had not recovered the exposed papyrus would have either been burned for land clearance for cultivation or would have decayed and ended up in the lake. Both of these outcomes would have resulted in further nutrient release into the lake and accelerated the eutrophication of the water. The lake drawdown had exposed a seed-bank, including papyrus and water hyacinth, on the lake sediment which germinated. As a result water hyacinth subsequently proliferated on the lake as the water level recovered (Nega et al., 2022). Also, as the lake levels rose the papyrus seedlings developed rhizomes close to the sediment surface which eventually detached from the sediment and floated as a mat on the surface of the water around the perimeter of the lake.

During drawdown there had been encroachment by local land users onto the exposed lake deposits of silt and peat. However, crops such as cocoyam (*Colocasia esculenta*) are far less efficient than the papyrus for nutrient removal from the polluted waters of the lake (Kansiime et al. 2007). Increasing population densities, economic development and agricultural encroachment along the shoreline of Lake Victoria are reducing the optimal performance of the papyrus dominated systems as carbon sinks. Furthermore, the added pressures associated with climate change, such as increasing temperatures and a changes pattern of rainfall may further negatively impact wetland functioning. It is imperative therefore that these multifunctional roles of papyrus wetlands form an integral part of future wetland management strategies and incorporate provisions for sustainable wetland use under future climate scenarios.

In Uganda there are examples of a hierarchy of councils and committees at five administrative levels varying from districts to village that propose policies which are informed by representatives of the local population. This decentralised structure has been used as a vehicle to train and equip district environmental officers to carry out wetland management functions at a local level which are implemented in a manner consistent with maintaining the livelihood of the local population (Maclean et al., 2014). Unfortunately, there are examples of development plans which focussed on gazetting of areas to assure long term conservation which led to conflicts between landowners and district government (Bikangaga et al. 2007). In these cases landowners in the area claimed that in the gazetting of the conservation area they were not consulted while others claimed that the gazetting was done arbitrarily and people lost access to land without compensation.

Although the freshwater wetlands that cover much of East and Central Africa are recognised as important resources for the people who live near them there is also, particularly at government level, a widespread view that alternative land uses have greater economic value. This has led to some relatively large government-backed agricultural schemes in which drainage of extensive swamp networks have been undertaken. Also, at a more local level numerous smaller wetlands have been drained or encroached upon. Satellite imagery has been used to monitor these changes and their observations suggest that more than 70% of the upland swamps in the Upper Nile and Congo basins had been lost in less than twenty years.

In Uganda, about 13% of the total land surface are wetlands and the government has recognised the contribution that wetlands make to the national economy and rural livelihood. It is estimated that over 50% of the wetlands are under some form of human use for agriculture or to provide materials for domestic use and the wetland sector is estimated to employ nearly 10% of the total population. In some cases policy makers have stimulated intensification of exploitation to improve livelihoods for riparian communities with the consequence is that, particularly on the shores of Lake Victoria, it has exceeded the potential for sustainable use which leads to degradation and loss of livelihood benefits for the riparian communities.

The Ugandan government recognised the contribution of wetlands to the National economy and rural livelihood but that these ecosystems were being threatened by population growth and increased pressure on natural resources. In 1989 they produced the National Wetland Strategic Plan followed in 1995 by the national policy for conservation and management of wetland resources. The plan attributed an economic value of the water treatment services supplied by Nakivubo wetlands, into which is discharged much of Kampala's waste, at US$1.7 million per year, and another US$100,000 for the wetland goods and products including crop cultivation, papyrus harvesting, brick making and fish farming (Kansiime et al., 2007). As a result, at government level there is a recognition that wetlands need to be conserved and contribute significantly to the National economy. This policy supports the concept of the 'wise use of wetlands' which depends on the maintenance of the ecological status of sustainable wetlands including papyrus swamps. However, there are often potential conflicts that emerge between some of these management options. For example, while conservation of the papyrus swamp ensures maintenance of the water treatment services, at the local level a subsistence farmer will prefer the papyrus removal in order to ensure the production of a food crop essential for the well-being of the local community. The message to policy makers is that they should avoid complacency about the possible adverse effects of papyrus removal, because there are often perceived disadvantages at the local level of conservation. These are very difficult dilemmas faced by many wetland development proposals (Bikangaga et al., 2007). There is no doubt that conservation brings advantages but all too often the disadvantages are not appreciated. On the other hand, despite evidence of serious ecological and environmental consequences, major wetland engineering schemes can be entered into which have the impact of devastating the wetlands that provide overlooked multiple ecosystem services.

Upper Nile Basin: Impacts of Altered Hydrology

At the same time as the increasing threat to papyrus wetlands from increasing population the wetlands have also been threatened by changes that have come from large scale schemes that alter the hydrology of these ecosystems These have largely focussed on the demand for water for the urban population and irrigation systems and the construction of dams primarily for the production of hydro-electricity. Clearly, one of the main problems in assessing the impacts on wetlands of development schemes is the absence of 'base-line' data providing information on conditions before and after change. One of the few examples of a serious attempt to do this was associated with the proposed construction of the Jonglei Canal in the Sudd wetland with a planned length of 350 km, twice as long as the Suez Canal. In 1988 Howell et al. published *The Jonglei Canal: Impact and Opportunity*, which aimed to present the scientific facts, the product of a number of different professional disciplines, together with an objective interpretation. Since the beginning of the last century there had been plans to divert the Nile from the swamps of South Sudan to save water lost to evaporation and transpiration and to improve the communications between the north and south. The Jonglei region consists of 11,000 km^2 of seasonally inundated flood plain. Preparation for the first serious attempt to dig the canal began in 1974 and the survey work of Howell et al. (1988) took place in 1977 before construction started in 1978. However, six years and 260 km later, this was brought to an abrupt halt by a local civil war. The planned development had always been the cause of suspicion and controversy, largely because of possible misunderstandings of the nature and extent of the local impact. It was also clear that the canal, once complete, would have significant implications for both the wildlife and local people whose migrations could be severely disrupted. This would have been the consequence of a reduced volume of water entering the Sudd and a predicted reduction of 20–80% of the swamp and 17–58% of the floodplain, depending on the actual annual river discharge from the Equatorial lakes, which were already known to exhibit unpredictable long-term variations. In addition, the incomplete canal continues to present a serious barrier to the movement of people, livestock and wildlife (Mitchell, 2013).

The impacts of ambitious plans by Uganda, and South Sudan for dams along the Bahr el Jebel flowing from Uganda and its tributaries are uncertain.

Although these upstream developments may not reduce the total flows of water through the wetlands, they will certainly have effects on the seasonal flows into the Sudd. This is because the spatial variations and seasonal dynamics, identified by the Satellite monitoring, which have emphasised the importance of the flood pulse to the Sudd wetland. Certainly, future water resource development in the region is likely to lead to disruption of the timing and magnitude of the flood pulse. However, more analysis is required to elucidate the details of spatial variation across these vast wetlands. The largest of these dams is the Karuma Hydroelectric Power Station which opened in September 2024. It is the largest power-generating installation in Uganda and is located on the Victoria Nile at the former location of the Karuma falls.

Impacts of Oil Extraction

The extraction of oil in the Upper Nile Region for several recent decades has had severe negative environmental and health impacts for wetlands and their occupants for a number of reasons. These include; oil spills and leakage that have contaminated water sources, habitat destruction as a result of clearing for development of oil infrastructure, contamination affecting vegetation growth and, for the local population, exposure to toxic chemicals that have led to a number of health issues. In addition, and probably most importantly, it has led to large-scale displacement of local populations and loss of livelihood affecting agriculture, pastoralism and fisheries.

Some of the most severe damage has been caused by the multinational exploitation of oil reserves discovered below the wetlands of the Sudd region and the Bahr el Ghazel. However, there appears to have been few rigorous assessments of the potential environmental impacts of oil extraction which was initiated in the 1980s, and those that were published were generally superficial and produced by organisation that had vested interests in the oil extraction industry. Unsurprisingly they concluded that there were no major issues that would give serious environmental concerns. This is despite the fact that any oil spillage into the Nile and the surrounding swamps would be anticipated to have significant environmental impact, particularly during the rainy season when the whole area is flooded. During this time any spillage would flow downstream northwards towards Khartoum and Cairo. Also, as the water recedes during periods of low flow, oil slicks would remain on pasture and in the fishing areas. Satellite imagery was also used to support the assertions that as there were few human inhabitants in these areas the direct impact on local populations would be limited. However, the images appear to have been carefully edited to ignore the evidence of movement of agri-pastoralists that travel seasonally with their herds. An Amnesty International document from 2006 reported that the exploitation of oilfields in southern Sudan, 'was accompanied by massed forced displacement and killings of the civilian population living in the area. Sudanese planes bombed villages and southern militia, supported by Sudanese armed forces, attacked villages killing people and destroying homes until the area was depopulated, in an apparent aim to clear the area of people for exploration and extraction', In fact the conflicts led to the displacement of tribes that had traditionally migrated through the wetland areas on an annual basis, and had forced them to abandon the nomadic life completely as chronicled in Richard Cockett's book *Sudan: The Failure and Division of an African State* (Cockett, 2016). When oil extraction began the exploration companies built drilling sites, pipelines and roadways on the tribe's grazing areas which blocked the migration routes of the cattle. Also, expatriate workers moved into the areas seeking economic opportunities, which put more competition for scarce water resources that had become contaminated with oil from the drilling sites that seeped into the local waterways. Meanwhile, oil production continues and now generates as much as 90% of South Sudan's government budget, although the government has been criticised for the weak enforcement of environmental regulations and a lack of transparency in the oil industry.

Further south in Uganda, since the early 2000s, a number of wells have been drilled by oil and gas operators on wetlands close to Lake Albert and within the Murchison Falls Conservation area and Uganda's oil industry has now reached development and production stages. The project has undergone environmental assessment and in 2019 an Environmental and Social Impact Assessment was approved by Uganda's National Environment Management Authority. The project involves the construction of a refinery with a capacity of 60,000 barrels of oil per day and the construction of a 1,443 km pipeline to transport oil from Uganda to the Tanzanian port of Tanga, on the coast of the Indian Ocean. The oil companies and Ugandan government are implementing measures to protect and conserve local biodiversity and ecosystems. However, concerns remain about the effectiveness of these.

Impacts of Invasive Plants

The Nile Basin wetlands are facing significant threats from invasive aquatic weeds that have been introduced by human activities in recent decades. The most prolific invasive plants appearing in the Nile Basin are the free floating macrophytes water hyacinth (*Eichornia crassipus*) and *Salvinia molesta*, both of which originate from South America but have become pan-tropical. Both species combine extensive asexual vegetative reproduction with potentially exponential population increase and horizontal mobility, by wind and currents, to show spectacular range of extension over water. Growth is dependent on a plentiful supply of nutrients under eutrophic conditions, when plants are able to double their numbers every two weeks (Batanouny and El-Fiky, 1975).

Water hyacinth has been recorded in Egypt since the 1850s, but it was the 1970s before it became an extensive problem (Batanouny and El-Fiky, 1975). It has now invaded most water bodies in the Nile Basin where it is creating severe adverse impacts on these water bodies' ecosystem functioning as well as causing problems with river navigation, which affects the daily life of the local population, while at the large scale it interrupts hydro power generation systems causing damage to turbines and interrupting power supply (Nega et al., 2022). Vast amounts of water hyacinth can cover open water, while wind and current action can compress them into a thick carpet; so thick that it has been reported that people can walk on them! The floating mats of water hyacinth reduce water currents and provide increased breeding grounds for malaria-causing *Anopheles* mosquito while the tangled roots are home for the snails that are a vector for the *Bilharzia* parasite. It has also been shown that the presence of water hyacinth significantly increases the rate of evapotranspiration from the water surface which reduces the amount of water flow downstream (Little, 1967). There are also numerous impacts on fishing activities including reducing access to the breeding, nursery and feeding grounds for some of the economically important fish species and therefore having a negative impact on the economic status of the fishery communities.

While water hyacinth is regarded as one of the world's most destructive weeds as it appears to be nearly impossible to control and eliminate due largely to its rapid growth rate, particularly in eutrophic waters (Rezania et al., 2015). Control methods to reduce water hyacinth invasions include biological and physical (manual and mechanical removal) approaches either singly or in combination. No single method has proved effective as each has its limitations and challenges but manual and mechanical removal has generally proved to be the most effective method (Rezania et al., 2015). Chemical control using herbicides is an immediate and short-term solution to remove water hyacinth but its use in aquatic systems where they can contaminate the whole ecosystems is not practical or environmentally acceptable. Biological control methods include releasing a range of organisms that potentially inhibit the growth of the water hyacinth. These include several insects, moths and fungal pathogens which have been shown to be effective under laboratory conditions. However, when released in the field they have not overwhelmed the growth of the water hyacinth, probably because of its very fast growth rate. Because none of the putative control methods are able to irradicate this invasive weed, there have been several proposals for utilising the plant when it is removed manually or mechanically. These include natural fertiliser substitutes, raw material for polymer and biopolymer production, protein sources for animal feed and feedstock for bioenergy either as briquettes, biogas or bioethanol production (Nega et al., 2022).

Salvinia molesta was first recognized as a rapid colonizer of eutrophic waters in East Africa on Lake Kariba after it was formed following the construction of the Kariba dam on the Zambezi River in 1958. By 1962 the fern covered 22% of the surface of Lake Kariba at its peak, corresponding to the period when the lake was eutrophic as a result of nutrients leached from flooded land as water levels rose (Marshall and Junor, 1981). In the following years the lake became less eutrophic as nutrients were lost through the outflow and there was increasing competition for nutrients from other ecosystem components, and by 1973 *Salvinia* covered only 5% of the lake. The decline was associated with the fall in nutrient availability but it also corresponded with the implementation of a biological control programme using the neotropical grasshopper *Paulinia acuminata*, which appears to have made a major contribution towards the decline in *Salvinia* (Marshall and Junor, 1981).

Salvinia appears to have been absent from the Nile Basin until 2017 when it was reported to be widespread on Lake Kyoga, Uganda (Andama et al., 2017). In 2019, its presence on the White Nile was first recorded in Sudan (Madani et al., 2019), and it has been suggested that the source of this invasion was probably Lake Kyoga. The floating fern was probably carried by the flow of the Bahr el Jebel, which winds through the papyrus swamps in the Sud region of South Sudan before flowing northwards to the White Nile. Although *Salvinia* can become an aggressive invasive under certain circumstances, as illustrated by its rapid expansion on Lake Kariba in the 1960s, the evidence from the Nile Basin is that currently it has local impacts but its regional significance is limited. Certainly, in the Nile Basin it clearly ranks second to water hyacinth as an aggressive alien aquatic plant.

Impacts of Ongoing Climate Change

Since its establishment in 1988, the International Panel on Climate Change (IPCC) has provided a series of objectives reviews of the current state of the art in climate change research, the latest of which is the sixth assessment report produced in 2022 (IPCC, 2022). Climate research and the ability to project future climate relies on conceptual computer models of the earth-atmosphere system in the form of global climate models (GCMs). As computing power increased these models evolved from simple energy balance-type models to fully coupled earth system models, downscaled to a specific river basin scale and referred to as regional climate models (RCMs) (Di Baldassarre et al., 2011).

The IPCC projections of future climate for Africa using RCMs indicate that temperatures will rise by 1.0–2.6°C by 2050, although there is low confidence in these predictions as a result of the role that land cover change such as deforestation and wetland decline, will have on regional climate (Junk et al., 2013). Rainfall projections suggest that by the end of the century the annual amount of precipitation in the Upper Nile basin could increase by up to 20% but at the same time dry season hot dry spells are projected to become more frequent. With the increasing demand for water from a rapidly increasing population that could double by the middle of this century, water stress in the region will increase. Compounding the seasonal drying is a projected increase in the year-to-year variation in rainfall, which may be due to a projected intensification of El Nino and La Nina cycles. Mitsch et al (2010) suggested that climate change could affect inland tropical wetlands in three distinct ways: (i) changes in the hydrologic pulses from upstream; (ii) changes in local precipitation patterns; and (iii) changes in temperature/humidity and subsequent evapotranspiration patterns. All three changes will occur in the Nile Basin but their combined effects are still uncertain (Di Baldassarre et al., 2011).

Indications of current changes in climate may be evident from the occurrence of irregular extreme rainfall events in the Lake Victoria catchment. For example, during late 2019 and early 2020, an unprecedented increase in the level of Lake Victoria occurred as a result of unusually heavy rainfall in the lake's catchment area and this led to massive flooding in the lake-adjacent areas. Although no single extreme climatic event is an indicator of climate change this stimulated the scientific debate on the potential impacts of a changing climate on Lake Victoria's water budget (Ogega et al., 2023) and the ecology of Lake Victoria. These events have, in particular, shown that there is an urgency in developing measures to ensure the future conservation of papyrus wetlands as well as meeting sustainable development goals for the provision of clean water, improved sanitation and the conservation of biodiversity around the lake.

Threats to wetlands that are exacerbated by climate change, particularly increased rainfall, include a widening variability in river flows, increased river siltation resulting from high rates of soil erosion in the basin, recurrent destructive floods in the low-lying areas, riparian land encroachment, eutrophication and the proliferation of invasive plants such as water hyacinth. Climate models of future climate in the region indicate that by the end of the century the amount of rain falling in the parts of the Lake Victoria Basin could increase by up to 25% in the eastern and 5–10% in the western part of the basin (Olaka et al., 2019). Climate projections were based on IPCC scenarios known as 'representative concentration pathways' (RCPs), portraying possible future greenhouse gas emission trajectories ranging from optimistic to pessimistic. At higher emission scenarios, the October to December season receives more rainfall than the March to May season. However the projections also show that as well as increasing annual rainfall there will be more extreme events of heat and dry spells which will severely impacts on vegetation as well as on the human populations of the region. Periods of heavy rainfall and subsequent runoff will reduce vegetation cover and increase soil erosion. Runoff will deposit sediments in waterways and wetlands as well as reducing the capacity of irrigation systems and hydropower generating systems. There is general agreement that the average annual temperatures of most tropical wetlands will increase. The predictions are that temperatures will rise by 1.0–2.6°C by 2050 (Junk et al., 2013) although uncertainties for the regional predictions are large as a result of the long-term climatic variability as well as the role of land-cover changes in modifying regional climate (Olaka et al., 2019). Annual discharges for rivers draining the Upper Nile Basin will vary considerably in response to changes in the climate in the coming decades and the amplitude of the oscillations in discharge will be very large after 2030 (Olaka et al., 2019). Recent climate modelling by Ogega et al. (2023) has suggested that more significant precipitation changes over the Lake Victoria Basin are likely to occur at the subseasonal level compared to the seasonal and annual scales, with a projected 47% increase in heavy precipitation events by the end of this century.

Adapting to Impacts of Climate Change

In attempts to adapt to the impacts of climate change there have been a number of efforts to develop policies and management strategies for the protection of wetlands in the Nile Basin. Olaka et al. (2019) conclude that climate adaptation strategies and policies should focus on managing the predicted high variability in river flow rates and coping with frequent severe droughts, floods, and pollution of waterways during floods. The retention of wetland vegetation, including papyrus swamps, will be important in regulating river flow and filtering pollutants from flood water.

The negative impacts related to climate change will be compounded by synergies with other stressors, including land use change and invasive species. This increases both the difficulties in maintaining and restoring wetlands and ultimately the risk of endemic species extinction. Di Baldassarre et al. (2011) suggest that while there is a need to better recognise and characterise the uncertainty of future climate change and its impacts on the ecology and hydrology of the Nile Basin there is also a requirement to go further to effectively support decision makers by proposing adaptation strategies and measures that have the potential to relieve the negative impacts.

The impact of changed hydrology, water availability and climate on the socioeconomic and political landscape of the region are therefore important to assess. Climate adaptation strategies and policies should focus strongly on managing the projected high variability in river flow rates and coping with frequent severe droughts, floods, and pollution of waterways during floods (Olaka et al., 2019). During extended droughts it has been predicted that, combined with increased populations, the changing climate could cause chronic water scarcity for more than 40% of the population. Competition for water could peak during drought years, resulting in conflicts between different water users in the river catchment. Water allocation and rights need to be carefully managed to minimise conflicts between communities, basins, economic sectors, and states. In particular, in the case of the Nile Basin, this involves avoiding conflicts between the downstream states of Egypt and Sudan as they try to assert control over the Nile flow, while the upstream states try to find ways to secure water for their own well-being. These problems of the fair distribution of the Nile's natural resources and the threats of 'water wars' in the region can be seen as ultimately the consequence of human impacts on the Nile Basin ecosystems.

The international body that is charged with the long term protection of the world's wetlands is the Ramsar Convention which is named after the Iranian city in which the treaty was signed. The overall objective of the Ramsar Convention, which was signed in 1971 and came into force in 1975, is 'the conservation and wise use of wetlands by national action and international cooperation as a means of achieving sustainable development throughout the world'. The wise use concept is understood by the Convention as 'the sustainable utilisation for the benefits of humankind in a way compatible with the maintenance of the natural properties of the ecosystem'. The convention originally had a narrow objective of conservation of wetlands as a habitat for birds. But it rapidly developed to cover all aspects of wetland conservation of biodiversity as well as the well-being of human communities. The wise use concept is understood by the convention as 'the sustainable utilisation for the benefits of humankind in a way compatible with the maintenance of the natural properties of the ecosystems'. Since 1975 the Ramsar Convention has provided an international policy framework for wetland management. However, as yet the Convention has not provided specific guidance on how objectives and targets for wetland conservation and management could be set and revised in the light of projected climate change (Moomaw et al., 2018).

National strategies have been developed in a number of Nile Basin countries aimed at adapting to projected changes in climate. These include the development of vulnerability and adaptation assessments, and the creation of technical and political institutes and their networking. Despite these developments more needs to be done, particularly in relation to promoting capacity development as a tool for supporting sustainable Nile Basin management. Part of the solution to understanding the impacts of change and protecting wetlands from permanent destruction will be through the development of accurate inventories that provide essential information including the type of wetland, its exact geographic location and area. The obvious solution is to gather data from satellite-based sensors (Jones et al., 2009). This earth observation data can be used to supplement national wetland inventories which vary greatly in the detail that has been obtained and the consistency of the observations. This information is essential for policy-makers and management. It can also be used to establish baselines against which the effectiveness of policies can be assessed. In the last few decades earth observation by satellites have had increasing capabilities in terms of spatial, temporal and spectral resolution which has allowed them to make affordable monitoring of the environment over time at global, regional and local scales. In 2003 The European Space Agency (ESA) in collaboration with the Ramsar Secretariat launched the 'GlobWetland' project in order to demonstrate the capabilities of Earth Observation technology to support inventorying, monitoring and assessment of wetland ecosystems. The ultimate intention was to provide wetland managers and national authorities with an efficient support tool for their management and conservation activities. For many areas in the Global South the GlobWetland data set represented the first and only updated digital geo-referenced maps over their wetlands. For example, the land cover analysis of the Lake Naivasha papyrus wetland in Kenya, covering a temporal range of 10 years, showed the significant impacts of increased agricultural encroachment around the lake. The GlobWetland project has demonstrated that Earth Observation technology now has the potential to support the efficient management of wetland areas through the collection of short-term and long-term data and information to allow the efficient inventory, assessment and monitoring of wetland sites and entire catchment areas, as well as the effective support for the development and implementation of restoration and rehabilitation plans (Jones et al. 2009). In addition wetlands feature strongly in the 'New Partnership for Africa's Development' (NEPAD). In the NEPAD 'Action Plan for the Environment' (2020) the aim is that 'African countries and their people have healthy and productive wetlands and watersheds that can support fundamental human needs including clean water, appropriate sanitation, food security and economic development'.

Conclusion

Papyrus wetlands currently face a myriad of threats to their survival, mainly from human activity which has for thousands of years first exploited it and then discarded it when it was no longer useful. In the River Nile catchment human impacts lead to papyrus becoming extinct with the loss of former Lower Nile wetlands and their conversion to irrigated croplands. Papyrus wetland destruction in the Upper Nile is now focussed on areas of high human population density or where the hydrological regime is no longer favourable for swamp survival. Populations are increasing rapidly, particularly around the perimeter of Lake Victoria, and it is here that papyrus is most threatened where land has been claimed for agriculture by drainage where possible or by encroachment, particularly when water levels fall seasonally. It is however an over simplification to relate papyrus decline exclusively to increased population density. Other reasons include: (i) the impact of invasive weeds, particularly water hyacinth, which inhibits the extension of floating papyrus mats across open water; (ii) seasonally fluctuating water levels which can strand papyrus in a drawdown and overwhelm the papyrus mat when water levels rise rapidly; (iii) ecological damage caused by pollution from oil exploration/extraction and runoff from adjacent land under intensive agriculture; and (iv) altered hydrology as a result of construction of dams and canals along the course of the Nile and its tributaries. The major impact of a changed future climate is projected to be increased variability in rainfall leading to rapid changes on water level alternating with periods of drought when water levels will fall. There are considerable disagreements between the predictions of different models, particularly in relation to the rate at which this change will occur, but it is possible that the resilience of papyrus wetlands may be exceeded this century by an unprecedented combination of climate change and associated disturbances. There is an urgent need to develop climate adaptation strategies and policies that may allow papyrus to survive this onslaught, but it will require effective collaboration between the Nile Basin states.

References

Andama, M., Ongom, R. and Lukubye, B. (2017). Proliferation of Salvinia molesta at Lake Kyoga landing sites as a result of anthropogenic influences. *Journal of Geoscience and Environment Protection*, 5, 160–173.

Basheer, M., Nechifor, V., Calzadilla, A., Gebrechorkos, S., Pritchard, D., Forsythe, N., Gonzalez, J.M., Sheffield, J., Fowler, H.J. and Harou, J.J. (2023). Cooperative adaptive management of the Nile River with climate and socio-economic uncertainties. *Nature Climate Change*, 13, 48–57.

Batanouny, K.H. and El-Fiky, A.M. (1975). The water hyacinth (*Eichhornia crassipes Solms*) in the Nile system, Egypt. *Aquatic Botany*, 1, 243–252.

Bikangaga, S., Picchi, M.P., Focardi, S., and Rossi, C. (2007). Perceived benefits of littoral wetlands in Uganda: A focus on the Nabugabo wetlands. *Wetlands Ecology and Management*, 15, 529–535.

Cockett, R. (2016) *The failure and division of an African state*. Second edition. New Haven, CT: Yale University Press,.

Di Baldassarre, G., Elshamy, M., van Griensven, A., Soliman, E., Kigobe, M., Ndomba, P., Mutemi, J., Mutua, F., Moges, S., Xuan, Y., Solomatine, D. and Uhlenbrook, S. (2011). Future hydrology and climate in the River Nile basin: A review. *Hydrological Sciences Journal–Journal des Sciences Hydrologiques*, 56, 199–211.

Foucault, A., and Stanley, D.J. (1989). Late Quaternary palaeoclimatic oscillations in East Africa recorded by heavy minerals in the Nile Delta. *Nature*, 339, 44–46.

Gaudet, J. (2014). *Papyrus: The plant that changed the world – from ancient Egypt to today's water wars*. New York: Pegasus Books.

Howell, P., Lock, M. and Cobbs, S. (eds). (1988). *The Jonglei Canal: Impact and opportunity*. Cambridge: Cambridge University Press.

IPCC. (2022). *Climate change 2022: Impacts, adaptation, and vulnerability*. Contribution of Working Group II to the Sixth Assessment Report of the Intergovernmental Panel on Climate Change. Cambridge: Cambridge University Press.

Isunju, J.B. and Kemp, J. (2016). Spatiotemporal analysis of encroachment on wetlands: a case of Nakivubo wetland in Kampala, Uganda. *Environmental Monitoring and Assessment*, 188, 1–17.

Jones, K., Lanthier, Y., van der Voet, P., van Valkengoed, E., Taylor, D., and Fernández-Prieto, D. (2009). Monitoring and assessment of wetlands using Earth Observation: The GlobWetland project. *Journal of Environmental Management*, 90, 2154–2169.

Junk, W.J., Bayley, P.B. and Sparks, R.E. (1989). The flood pulse concept in river-floodplain systems. *Canadian Special Publication of Fisheries and Aquatic Sciences*, 106, 110–127.

Junk, W.J., An, S., Finlayson, C.M., Gopal, B., Květ, J., Mitchell, S.A., Mitsch, W.J. and Robarts, R.D. (2013). Current state of knowledge regarding the world's wetlands and their future under global climate change: A synthesis. *Aquatic Sciences*, 75, 151–167.

Kansiime, F., Saunders, M. J. and Loiselle, S. A. (2007). Functioning and dynamics of wetland vegetation of Lake Victoria: An overview. *Wetlands Ecology and Management*, 15, 443–451.

Kansiime, F., Kateyo, E., Oryem-Origa, H., and Mucunguzi, P. (2007). Nutrient status and retention in pristine and disturbed wetlands in Uganda: Management implications. *Wetlands Ecology and Management*, 15, 453–467.

Kiwango, Y. A. and Wolanski, E. (2008). Papyrus wetlands, nutrient balance, fisheries collapse, food security, and Lake Victoria level decline in 2000–2006. *Wetland Ecology & Management*, 16, 89–96.

Little, E.C.S. (1967). Progress report on transpiration of some tropical water weeds. *International Journal of Pest Management C*, 13(2), 127–132.

Maclean, I.M.D., Bird, J.P. and Hassall, M. (2014). Papyrus swamp drainage and the conservation status of their avifauna. *Wetlands Ecology and Management*, 22, 115–127.

Madani, I., Sinada, F. and Mohammed, T. (2019). Salvinia molesta (Salviniaceae) a new record for the flora of Sudan. *International Journal of Science, Environment and Technology*, 8, 132–140.

Manning, J.G., Ludlow, F., Stine, A.R., Boos, W.R., Sigl, M. and Marlon, J.R. (2017). Volcanic suppression of Nile summer flooding triggers revolt and constrains inter-state conflict in ancient Egypt. *Nature communications*, 8, 900.

Marshall, B.E. and Junor, F.J.R. (1981). The decline of *Salvinia molesta* on Lake Kariba. *Hydrobiologia*, 83, 477–484.

Mitchell, S.A. (2013). The status of wetlands, threats and the predicted effect of global climate change: The situation in Sub-Saharan Africa. *Aquatic Sciences*, 75, 95–112.

Mitsch, W. J., Nahlik, A., Wolski, P., Bernal, B., Zhang, L. and Ramberg, L. (2010). Tropical wetlands: Seasonal hydrologic pulsing, carbon sequestration, and methane emissions. *Wetlands Ecology and Management*, 18, 573–586.

Moomaw, W.R., Chmura, G.L., Davies, G.T., Finlayson, C.M., Middleton, B.A., Natali, S.M., Perry, J.E., Roulet, N. and Sutton-Grier, A.E. (2018). Wetlands in a changing climate: science, policy and management. *Wetlands*, 38, 183–205.

Nega, D.T., Ramayya, A.V., Afessa, M.M. and Manenti, F. (2022). Invasive water hyacinth challenges, opportunities, mitigation, and policy implications: The case of the Nile Basin. In G. Shukla, J.A. Bhat, S. Chakravarty, A.W. Almutairi & M. Li (eds), *Floristic diversity: Biology and conservation*. London: IntechOpen.

Ngodhe, S.O., Nyamai, D. and Owato, G. (2016). Detecting changes, causes and future prospects of Kenyan wetlands and their conservation. *International Journal of Fisheries and Aquatic Studies*, 4, 24–29.

Odada, E.O., Olago, D.O., Kulindwa, K., Ntiba, M. and Wandiga, S. (2004). Mitigation of environmental problems in Lake Victoria, East Africa: Causal chain and policy options analyses. *Ambio: A Journal of the Human Environment*, 33, 13–23.

Ogega, O.M., Scoccimarro, E., Misiani, H. and Mbugua, J. (2023). Extreme climatic events to intensify over the Lake Victoria Basin under global warming. *Scientific Reports*, 13, 9729.

Olaka, L.A., Ogutu, J.O., Said, M.Y. and Oludhe, C. (2019). Projected climatic and hydrologic changes to Lake Victoria Basin Rivers under three RCP emission scenarios for 2015–2100 and impacts on the water sector. *Water*, 11, 1449.

Ondiek, R.A., Vuolo, F., Kipkemboi, J., Kitaka, N., Lautsch, E., Hein, T. and Schmid, E. (2020). Socio-economic determinants of land use/cover change in wetlands in East Africa: A case study analysis of the Anyiko Wetland, Kenya. *Frontiers in Environmental Science*, 7, 207. doi:10.3389/fenvs.201900207.

Owino, A.O. and Ryan, P.G. (2007). Recent papyrus swamp habitat loss and conservation implications in western Kenya. *Wetlands Ecology and Management*, 15, 1–12.

Pietroiusti, R., Vanderkelen, I., Otto, F.E., Barnes, C., Temple, L., Akurut, M., Bally, P., van Lipzig, N.P. and Thiery, W. (2024). Possible role of anthropogenic climate change in the record-breaking 2020 Lake Victoria levels and floods. *Earth System Dynamics*, 15, 225–264.

Ramsar Convention Secretariat. (2007). Wise use of wetlands: A conceptual framework for the wise use of wetlands. In *Ramsar handbooks for the wise use of wetlands* (3rd ed., Vol. 1). Ramsar Convention Secretariat.

Rezania, S., Ponraj, M., Din, M.F.M., Songip, A.R., Sairan, F.M. and Chelliapan, S. (2015). The diverse applications of water hyacinth with main focus on sustainable energy and production for new era: An overview. *Renewable and Sustainable Energy Reviews*, 41, 943–954.

Said, R. (1993). *The River Nile: Geology, hydrology and utilization*. Oxford: Pergamon.

Schuyt, K.D. (2005). Economic consequences of wetland degradation for local populations in Africa. *Ecological Economics*, 53, 177–190.

Terer, T., Muasya, A.M., Higgins, S., Gaudet, J.J. and Triest, L. (2014). Importance of seedling recruitment for regeneration and maintaining genetic diversity of *Cyperus papyrus* during drawdown in Lake Naivasha, Kenya. *Aquatic Botany*, 116, 93–102.

Turyahabwe, N., Tumusiime, D. M., Kakuru, W. and Barasa, B. (2013). Wetland use/cover changes and local perception in Uganda. *Sustainable Agriculture Research*, 2, 95–105.

van Dam, A. A., Kipkemboi, J., Mazvimavi, D. and Irvine, K. (2014). A synthesis of past, current and future research for protection and management of papyrus (*Cyperus papyrus* L.) wetland in Africa. *Wetland Ecology and Management*, 22, 99–114.

Verschuren, D., Johnson, T.C., Kling, H.J., Edgington, D.N., Leavitt, P.R., Brown, E.T., Talbot, M.R. and Hecky, R.E. (2002). History and timing of human impact on Lake Victoria, East Africa. *Proceedings of the Royal Society of London. Series B: Biological Sciences*, 269(1488), 289–294.

Williams, M. (2019). *Nile Basin: Quaternary geology, geomorphology and prehistoric environments*. Cambridge: Cambridge University Press.

Willbroad, B. and Kiyawa, S.A. (2019). Sustainable management and conservation of wetland resources in Uganda: A review. *Journal of Environment and Health Science*, 5, 47–51.

Zhao, X., Liu, Y., Thomas, I., Salem, A., Wang, W., Alassal, S.E., Jiang, F., Sun, Q., Chen, J., Finlayson, B., Wilson, P. and Chen, Z. (2022). Herding then farming in the Nile Delta. *Communications Earth & Environment*, 3, 88. doi:10.1038/s43247-022-00416-7.

6 Ecosystem Services and Sustainable Management of Papyrus Wetlands

Introduction

Freshwater wetlands are integral elements of river basins throughout the world and they play an essential role in regulating the hydrological systems that the river basins contain. At the same time, they are extremely important resources for the people who live near them (see Chapter 2). However, in many cases the encroachment onto the wetlands and drainage for the cultivation of food crops have had serious impacts upon the regulatory functions of the wetlands. In tropical wetlands, drainage and cultivation is often initiated in response to local food shortages due to pressure from increasing population density. Population growth in communities that have traditionally cultivated the dryland margins of wetlands for crop production has stimulated the encroachment on to the margins of the wetlands. However, more intensive forms of use, which have involved drainage of whole wetlands and multi-cropping on the exposed soils, have had much larger impacts. This land use is beyond subsistence agriculture and offers the opportunity for additional income generation through the cultivation of locally marketable crops such as vegetables and sugar cane (Dixon and Wood, 2003). There is, however, a need to develop wetland management practices which minimise the alteration of the hydrological regime while retaining the maximum range of benefits for wetland-dependent local communities. This is because although wetland drainage and cultivation can accrue benefits in the short term there are concerns about the sustainability of these benefits because of the long term degradation of the land. The land use change from pristine wetlands to intensive agricultural production has major implications for freshwater wetland management and in areas where this form of cultivation is an essential part of livelihood there is a need to develop wetland management practices that limit the negative impacts on the hydrological regime and maximise the wide range of additional benefits that wetlands can provide. Management and decision-making for wetlands therefore needs an integrated approach in which all the potential benefits to human society deriving from the functioning of ecosystems are identified, their

DOI: 10.4324/9781003131540-6

importance is assessed and objectives about their desired outputs are formulated. Broadly speaking these benefits can be classified as 'ecosystem services' and in this chapter we review how the concepts of ecosystem services have developed and how they can be applied in attempting to achieve the sustainable development of papyrus wetlands.

Identifying Ecosystem Services

Ecosystem services are the many and varied benefits to society gifted by the natural environment. They are the natural functions of ecosystems which are directly or indirectly used or appropriated by human society. Ecosystem services are provided by the 'natural capital' of the ecosystem. Natural capital refers to the living and non-living components of ecosystems that contribute to the generation of goods and services of value for society. It can be defined as the stocks of natural assets which include geology, soil, air, water and all living things. It is from this natural capital that society derives the ecosystem services. The distinction between ecosystem services and natural capital is that ecosystem services are the flows of benefits which are gained from natural ecosystems, and natural capital is the stock of natural ecosystems from which these benefits flow.

Identification and quantification of the ecosystem services provided by natural environments can be a powerful tool to inform policy makers and to aid conservation (Langan et al., 2018, Xu et al., 2018). The concept of 'ecosystem services' emerged during the 1970s and has been increasingly developed and refined in the subsequent decades. This development was stimulated by the outcomes from the Millennium Ecosystem Assessment (MEA, 2005) which was a major UN-sponsored review of the impact of human actions on ecosystems and human well-being. The development of the MEA had its roots in the increasing demand by policy-makers to transform research findings that had focused on identifying and quantifying ecosystem services and transforming them into actual management practices and allowing the sustainable management of diverse ecosystems. Furthermore, it allowed the identification of the associated trade-offs between alternative management options.

The Millenium Ecosystem Assessment identified four major categories of ecosystem services: provisioning, regulating, support and cultural. Common ecosystem services provided by ecosystems include:

- Provisioning services
    - *food, freshwater, fibre* and *fuel.*
- Regulating services
    - *climate regulation, water purification, waste water treatment, erosion* and *flood control.*

- Supporting services
  - *primary production, nutrient cycling* and *soil formation.*
- Cultural services
  - *recreation, aesthetics, education, spiritual* and *religious* values.

(Xu et al., 2018, Thorslund et al., 2017)

Global analyses of ecosystem services have shown that wetlands are among the world's most important ecosystems in terms of assets as they provide multiple ecosystem services for people and nature. These include water supply and purification, climate regulation, flood and drought mitigation, food, fibre and fuel as well as livelihoods. Costanza et al. (2014) estimated that wetlands account for approximately 23.2% of the total value of global ecosystem services. However, at the same time they are fragile ecosystems which are very vulnerable to damage, particularly as a result of over exploitation by humans as well as the predicted negative impacts of future climate change. It is estimated that between 1970 and 2015, inland and coastal wetlands both declined by about 35% worldwide. That is three times the rate of global forest loss. Consequently, while wetlands supply a wide range of ecosystem services they remain one of the most threatened ecosystems globally.

Because wetlands are increasingly threatened, policies need to be put in place to protect them while continuing to allow access to local communities. Implementation of these policies may need to be agreed and regulated by decisions made at the highest levels of government. For example, in Uganda, where it is known that several million people in riparian communities are provided with wetland ecosystem services that include water, fish, construction materials, food crops and livestock grazing, these benefits have been recognised by the Ugandan 1995 constitution which states that 'wetlands … shall be held in trust by the government for the good of all people'.

In this chapter the focus is on the ecosystem services provided by the extensive papyrus wetlands of the River Nile Basin and their sustainable use. Some of the wetland areas associated with the lower Nile were occupied by humans for millennia and they provided important support for the local populations throughout that time. During that time the interactions between the human population and the papyrus swamps maintained them in a pristine state using what we might refer to as 'sustainable management' (Morrison et al., 2012). In fact, the wide range of uses the ancient Egyptian dynasties found for papyrus over many hundreds of years must attest to the use of papyrus wetlands as a long term sustainable source of raw material.

In recent years the potential for sustainable use of the swamps has been increasingly recognised by international development agencies who have channelled significant funding in helping to exploit these wetlands in a sustainable fashion. This development has been based on research with the intention of improving the understanding of how papyrus wetlands function and how they can be retained or even expanded. Much of this development research has been focussed on Uganda and around Lake Victoria (Junk et al. 2013, Mitchell 2013, van Dam et al., 2014). As a result, Uganda has led the way in incorporating this knowledge into policy and management with the aim of providing sustainable livelihoods for people living close to wetlands (Namaalwa et al., 2013, Langan et al., 2019).

Ecosystem Services of Papyrus Swamps

Since the term ecosystem service is now internationally accepted as a description of how ecosystems sustain and fulfil human life, it is important to establish how its four major categories of services have been applied to papyrus wetlands:

Provisioning Services

As we have seen, a provisioning service is any type of benefit to people that can be extracted from an ecosystem. These benefits are tangible products including food, fibre, water, and energy. Provisioning services in the more developed economies frequently have well-developed markets and valuation systems. However, in tropical wetlands these provisioning services can be difficult to evaluate, largely because the services they provide are utilised locally and do not enter into the conventional markets (Rebelo et al., 2010). Market availability can also depend on location and the varying demand for the service. For example, during the time of ancient Egypt, papyrus culms provided higher relative valuation than they do today. At that time the principal demand for papyrus was as raw material for the manufacture of writing material in the form of papyrus rolls. There were also additional marketable uses for the papyrus, including ropes, baskets, roof thatching and matts for floor covering and other household goods.

The supplies of provisioning services are dependent on the primary productivity of the ecosystem. In the case of largely pristine papyrus wetlands the very high productivity of the plant (Saunders et al., 2007) has the potential to provide, very efficiently, large amounts of plant biomass on a sustainable basis. Today, many rural communities living in close proximity to papyrus wetlands rely to a greater or lesser extent on the availability of the papyrus plant biomass to support their livelihood. Potentially, virtually every part of the papyrus plant, from rhizome to umbel, can be utilised. This growing stock of living material is the 'natural capital' of the wetland that has the potential to provide fuel from rhizomes and dried culms as well as a range of marketable goods including furniture, ropes, mats, baskets and other handcrafts (Morrison et al., 2012). Papyrus wetlands are also important for supporting commercial fishing, as their rhizome systems provide a protective habitat for some species of fish (Kiwango and Wolanski, 2008) as well as a breeding environment and shelter for fish that subsequently move to adjacent river and lake fisheries. When papyrus swamps are encroached on, or are completely removed for crop production, the land continues to provide a provisioning service in the form of the crops that have a marketable value. In this case the relative values of the ecosystem services can be compared in the form of cost-benefit comparisons between the provisioning services of marketable papyrus and the value of crops grown on the swamp area.

Regulating Services

Regulating services are the benefits provided by ecosystem processes that moderate natural phenomena such as photosynthesis and nutrient uptake. The regulating services control the states and rates of physical and biotic systems and processes in ways that are beneficial to humans and maintain environmental condition favourable to life. Regulating services include water purification, erosion and flood control, carbon storage and local climate regulation.

Two particularly important regulating services in wetlands are carbon storage and waste-water purification. Carbon storage or sequestration is the process involved in the capture of CO_2 fixed in photosynthesis and long term storage of the carbon in natural sinks, with minimal chance of the CO_2 being released back to the atmosphere. It involves the increase in stocks of resilient carbon in the ecosystem, either in biomass or the peat formed from dead and decayed papyrus. The carbon sequestration potential of a wetland is determined by the rate of carbon storage and the maximum amount of carbon that can be retained in the ecosystem (see Chapter 4).

There is increasing evidence that the control of global warming cannot be achieved by reducing atmospheric CO_2 emissions alone, and will require additional sequestering of significant amounts of CO_2 currently in the atmosphere. Natural ecosystems, and particularly wetlands, are now recognised as one of the most important, cost-effective and efficient options for sequestering significant amounts of atmospheric CO_2. As a consequence of the high productivity rates of papyrus, and slow decomposition of dead and decaying vegetation in the anaerobic water of the rhizomes and roots, papyrus wetlands can potentially achieve high rates of carbon accumulation in the form of partly decomposed biomass, or peat (Saunders et al., 2014).

The water quality ecosystem service in papyrus swamps is regulated by the processes of sedimentation, filtration, physical and chemical immobilisation, microbial interactions and uptake by vegetation as the effluent enters and moves through the swamp. Water purification in swamps occurs when the roots and rhizomes intercept and trap nutrients entering water bodies from the surrounding catchment. The effectiveness of these processes is dependent on the control of water flow through the swamp as well as the width and continuity of the fringing vegetation. In many locations the wetlands are now almost wholly responsible for cleaning up polluted effluents from heavily populated areas. For example, in Uganda, sewage from 40% of the residents of the city of Kampala, with a population of almost two million, is discharged into the 5.3 km^2 Nakivubo wetland. During the passage of the effluent through the wetland, it has been demonstrated that the papyrus vegetation absorbs nutrients and pollutants are reduced markedly before the water enters Lake Victoria (Kansiime et al., 2007). This avoids the contamination of the lake which is the principal water source supplying Kampala.

A fundamental role of papyrus swamps in the ecotone between the dry land and open water is to buffer the runoff of water from the land. The wetland can store excess water in the rainy season and slowly release it to the ground water, streams and rivers during the dry season. During the movement of water through the swamp it is a 'sink' for nitrogen and phosphorus contained in the runoff. These nutrients are taken up by the roots of papyrus and are converted from largely inorganic to organic forms. The productivity of the papyrus vegetation is dependent on the supply of nutrients from the land as photosynthesis is generally limited by the supply of either nitrogen or phosphorus. The main processes underlying nitrogen and phosphorus retention are complex and involve accumulation and storage in organic matter after uptake by the papyrus roots and adsorption onto detritus and sediment after senescence, death and decomposition of the plant (Gaudet and Muthuri, 1981; Pacini et al., 2018; Hes et al. 2014). In model studies nitrogen and phosphorus retention was estimated to be more than 40% for both nutrients when the papyrus is harvested regularly but only 7% for nitrogen and 4% for phosphorus when the natural cycle of senescence occurs (Hes and van Dam, 2019). The reason for higher retention in harvested papyrus was the stimulation in growth rate of new shoots after harvesting and nutrient uptake.

Supporting Services

Supporting services are the underlying ecological processes that enable the ecosystem to function and are the services that are necessary for the production of all other ecosystem services. Important supporting services in papyrus swamps are nutrient cycling, primary production as the net balance between photosynthesis and respiration, and the water cycle. Ecosystems could not survive without the consistent support of these underlying natural processes which maintain the ecosystem integrity, functioning and resilience and enable the ecosystem to produce the more directly exploited provisioning ecosystem services. Other supporting services are water cycling (see Chapter 4), soil formation and retention, and the maintenance of biodiversity.

Nutrient cycling in wetlands is the process by which nutrients are exchanged and transformed within the ecosystem. The cycling of nutrients involves uptake by vegetation from the water or sediment, storage and metabolism in the plant biomass and, on senescence and death of the vegetation, the release of nutrients back to the water or storage in accumulating dead matter. Nutrient cycling can improve water quality by removing excess nutrients from the water as well as support biodiversity by providing a nutrient rich environment for plants and animals and maintain ecosystem productivity by regulating nutrient availability. Human activity can disrupt the nutrient cycles, particularly of nitrogen and phosphorus, by increased release from agricultural runoff and sewage. This nutrient enrichment of the water causes eutrophication and water quality degradation. Hes and van Dam (2019) have developed numerical models of nitrogen and phosphorus cycling and retention in papyrus dominated wetlands to quantify the effects of wetland vegetation in regulating the processes driving the nitrogen and phosphorus nutrient cycles.

Cultural Services

Cultural ecosystem services are the non-material benefits that people obtain from ecosystems through recreation, tourism, intellectual development, spiritual development, reflection and creative and aesthetic experiences. Papyrus wetlands were widely used in ancient Egypt as sacred places for worship and as a source of myths and traditional wisdom and socio-cultural uses of papyrus are still present among most populations living close to Nile Basin wetlands (van Dam and Kipkemboi, 2016). For example, the Nyando wetland in western Kenya plays an important role in the tradition of the Luo community, who use the wetland for spiritual cleansing and rituals of sending away evil spirits. Papyrus is also an important component of traditional rituals and dancing during cleaning and blessing ceremonies practised by the Endoroi tribe living around Loboi Swamp, on the shores of Lake Bogoria, Kenya and the Endoroi name given to papyrus translates into 'crown of majesty'. Perceptions of the value of cultural ecosystem services will differ between individuals and communities and are likely to be locally specific and as a result cultural service assessments are likely to be mainly arbitrary and limited in scope. However, one marketable cultural service that has been developed relatively recently is eco-tourism, which is encouraged by the presence of papyrus specialist bird species (Rongoei et al. 2013).

Biodiversity and Trade-offs between Ecosystem Services

Biodiversity is a foundation for ecosystem services as it supports regulating services, particularly climate regulation by influencing carbon sequestration and water cycle regulation by influencing evapotranspiration. As papyrus, particularly in permanently flooded areas, grows in dense monospecific stands that exclude other emergent vegetation the papyrus swamp ecosystem appears, at first sight, to exhibit low levels of biodiversity. However this is not the case as the swamps support large numbers of fish, amphibians, reptiles, mammals and bird species that live in them. In fact, intact papyrus swamps are extremely important for wildlife conservation in general but they are particularly significant as the host of a suite of papyrus specialist birds, including the papyrus yellow warbler, papyrus gonolek (*Laniarius mufumbiri*), white-winged warbler (*Bradypterus carpalis*), papyrus canary (*Serinus koliensis*) and Carruthers's cisticola (*Cisticola carruthersi*). The papyrus yellow warbler and papyrus gonolek are both listed as globally threatened and require urgent actions to conserve papyrus swamps to protect their populations (Maclean et al., 2006). Wetlands management is therefore faced with potential conflicts between human exploitation of natural resources and conservation. This may involve restricting access to designated areas in order to prevent harvesting of the papyrus canopy where the birds reside. However, it now appears that most of the bird species are in fact tolerant of low intensity disturbance, such as limited level papyrus harvesting, and it has been suggested that policy makers need not totally prohibit harvesting of papyrus or other disturbance of the swamp in designated areas to conserve biodiversity (Donaldson et al., 2016). It has been proposed that this could generate good will locally and encourage rural householders to comply with this approach to conservation policy. The conclusion is that attempts to conserve wetlands, and the biodiversity dependent on them, does not necessarily involve the complete exclusion of local people who rely on resources the swamps provide. These conclusions highlight the potential to amicably combine subsistence resource use and conservation management in papyrus wetlands (Donaldson et al., 2016). The evidence is that biodiversity in papyrus wetlands has multiple roles in the delivery of ecosystem services; as a regulator of ecosystem processes, as a service in itself and as a good, and it is argued that effective ecosystem management probably depends on both the optimisation of ecosystem service delivery and the conservation of biodiversity (Mace et al., 2012). Bringing these together will be essential as it becomes clear that most ecosystems must be used for multiple services and this requires trade-offs between ecosystem services. Trade-offs between ecosystem services arise when the provisions of one ecosystem service are reduced as a consequence of increased use of another ecosystem service. Trade-off between ecosystem services are management choices which can change the type, magnitude and relative mix of services provided by the ecosystem. Understanding these trade-offs between processes providing services is key to improving guidelines for wetland management (Swallow et al., 2009).

Temporal Changes in Ecosystem Services

The ecosystem services of papyrus wetlands are usually regionally and locally specific as they are driven by the local social and economic situation and the demand for specific provisioning services. They depend on the one hand on the natural features and functions of the ecosystems, and on the other, their use by different stakeholders. It is likely that the ecosystem services provided by papyrus wetlands have changed as the wetlands have been placed under pressure from increasing population and encroachment for exploitation of their services (Dixon and Wood, 2003). An example of this comes from an area of wetlands considered to be a typical example of exploited papyrus wetlands in the Lake Victoria Basin that have formed in the Nyando River floodplain as it enters the Nyanza Gulf in the eastern region of Lake Victoria. The three ecological zones in the wetlands are permanent swamps, seasonal swamps and floodplains of the Nyando River. Rongoei et al. (2013) used semi-structured community interviews to obtain local knowledge of the wetland and its changes while open-ended questions were asked to recall changes that had occurred in the wetlands within the previous 50 years. This information was supplemented with results from research published in the scientific literature and government reports to focus on historical to present day knowledge on wetland resource use and provisioning ecosystem services. The results show that the main drivers of change over the 50 year period were hydrology and population increases associated with changes in livelihood activities. The major hydrological event that shaped the Nyando wetlands was the heavy rains of 1961/1962 that affected most parts of the Lake Victoria basin. This rain led to a rise in Lake Victoria levels that resulted in extensive inundation along the lake margins and the expansion of the existing band of papyrus vegetation to more extensive fringing wetlands, that led to a loss of cropland. People's livelihood activities at the time included fishing, mainly in the wet season, and livestock herding and seasonal wetland cultivation in the dry season (Rongoei et al., 2013). However, since 1961, seasonal patterns of rainfall and drought have been less predictable and there have been abnormal floods in some years and droughts in others. Between 1998 and 2005 there was a prolonged drought and the lake level declined by 2.3 m and has not, until recently, returned to the 1961/1962 levels. The increasing population during this time resulted in increased pressure on the fisheries, leading to overfishing and fish catch decline, as well as increased necessity for farming and a demand for other provisioning services including papyrus harvesting. Before 1980, activities of farmers were spatially separated in the terrestrial and wetland ecosystems, depending on the on the season, with fishing in the lake during the dry season and farming or livestock herding in the rainy season. Since then, increasing population pressures and changes in hydrological regimes have increased demand for provisioning services in the form of wetland farming. Papyrus harvesting has also increased since the 1990s to become an important provisioning service for increasing household income. Indeed it may become the only income earner when both dry and wetland farming have failed as a result of either drought or flooding events. Rongoei et al. (2013) concluded that the maximising of provisioning services in this wetland has led to a decline in provisioning services as a result of a decline in biodiversity and nutrient and sediment retention leading to the long term degradation of the papyrus wetland ecosystem.

Valuing Ecosystem Services

It has frequently been argued that although the idea of identifying the range of ecosystem services may be simple in concept their use in making management and policy decisions is complex because of the difficulty in applying a meaningful value to the service. Although valuing ecosystem services provided by the natural environment is potentially a powerful tool to aid conservation and sustainable utilisation of wetlands there are a number of uncertainties and caveats in relation to the implementation of the process. These include the fact that the almost universally used monitory valuation process is automatically weighted towards countries with strong currencies and high GDP, and it conversely undervalues ecosystems in low-income countries, and furthermore the valuation can apply at different scales, from the local to the global scale (Langan et al., 2019; Xu et al., 2018).

A major issue in assessing and interpreting the value of ecosystem services in wetlands has been the lack of common terminology. In order to address this issue, Turner et al. (2000) proposed an integrated wetland research framework which involves a combination of economic evaluation, integrated modelling, stake-holder analysis and multicriteria evaluation. It is suggested that to make progress in the important work of integrating models, natural and social science researchers should reach agreement on the following: (i) terminology and typology appropriate to the valuation, (ii) the scale of effects to be analysed, (iii) valuation methodologies, (iv) links between valuation and systems scenario analysis, (v) the transferability of information and results in both scientific and economic realm, (vi) the focus of the analytical approach, whether thematic or by site, and (vii) consideration of valuation within the context of the prevailing political and social framework.

The value of ecosystems can be divided into two categories; economic value and intrinsic value. The intrinsic value of a wetland reflects the view that nature has a value in its own right, independent of its human use. This is possibly highly controversial and difficult to incorporate into decision-making processes. Therefore, most analyses for wetlands focus primarily on the economic trade-offs of using, conserving or reclaiming the wetlands (Maclean et al., 2011). The economic values can be divided into direct use values gained from direct utilisation of the ecosystem, and indirect use-values are defined as those non-monetary goods and services derived from the functioning and existence of the ecosystem.

A difficulty with integrating information about the value of these ecosystem services is that they are often realised at different spatial scales (Thorslund et al. 2017). For example, regulating services require integration of ecological and socio-economic studies which are carried out at a plot scale in the case of the former and community scale in the case of the latter. These in turn need to be integrated with remote sensing data which can be at a much larger spatial scale. This integration can only be achieved by the development of simulation models which account for the whole range of ecosystem services provided by the wetland. The difficulty in using this information in a decision-making process is that the benefits for the 'direct use' local end-users is in conflict with the potentially much larger benefits to the indirect users who are generally dispersed much more widely across the catchment area for the rivers supplying the wetland. For example, at a regional scale one of the most significant activities that impacts on wetlands is the impoundment for dams and irrigation schemes (Schuyt, 2005). A consequence of establishing these is that they redirect the benefits of the dam away from the indigenous population to deliver ecosystem services to distant populations. This means more assured supply of benefits of water supply for many but depleted wetland services for the communities local to the wetlands. For the local communities the impoundment has the effect of reducing the area flooded and therefore reducing the floodplain productivity while the process also traps sediments and associated nutrients, starving the downstream areas of new soils and essential nutrients. This illustrates how irrigation schemes, established by both local governments and at larger scales by international agri-business consortia, both fail to support the local population and may be less sustainable in the long term.

Ecosystem Services from Papyrus Wetlands

Since the publication of the Millennium Ecosystem Assessment (MEA, 2005) there have been many published reviews that have assessed the impacts of current or changed management of wetlands on the provision of ecosystem services, but the vast majority of these have focussed on wetlands in North America, Europe, China and Australia, while sub-Sahara has been poorly represented (Xu et al., 2020). However, there have been a small number of reviews of the value of the Nile Basin papyrus wetlands in providing beneficial ecosystem services to the riparian populations close to the wetlands and to the wider community, and sustaining the services by using suitable management (Morrison et al., 2012; van Dam et al., 2016; Langan et al., 2019). Values have been expressed in different ways, including in qualitative and monetary terms. Clearly, management and decision making for wetlands needs an integrated approach, in which all ecosystem services are identified and their importance assessed. Ultimately, for the most effective management of the ecosystem there is a need to optimise the presentation of the values of the ecosystem services for the different beneficiaries of the services locally and regionally.

Unfortunately, in practice there is a lack of information on many of the characteristics of papyrus wetlands required to furnish detailed analyses of ecosystem services (Langan et al., 2018). For example, provisioning services often get priority over regulating services despite the fact that regulating services probably have a higher monetary value. However, regulating services are difficult to value because of the lack of knowledge of their underlying processes so their true value not known and is not incorporated in the analysis. Despite these difficulties, there have been a number of attempts to incorporate the concepts using the limited amount of information that is available. One of the first economic valuations of ecosystem services of papyrus wetland was made by Emerton et al. (1998) on the Nakivubo urban wetland, close to Kampala, the capital of Uganda. The Nakivubo wetland is approximately 5.29 km^2 in area, has a total catchment extending over 40 km^2, and is fed by the Nakivubo River that runs from Kampala's central industrial district and enters Lake Victoria at Murchison Bay, after passing through dense residential settlements and commercial areas. Due to the wetland's geographic position close to Kampala, it acts as an important sink for domestic and industrial waste of the city, but it is also encroached on by the rapidly increasing local population for cultivation of subsistence crops as well as industrial and residential expansion. In the economic valuation study of Emerton et al. (1998), four provisioning services and one regulating service were quantified. The provisioning services were, crop cultivation, papyrus harvesting, brick making and fish farming, while the regulating service was water treatment and purification. There were water-logged and drier areas of the swamp which were cultivated differently. In the permanently water-logged area of the swamp mainly cocoyams (*Colocasia esculenta*) and sugar-cane were grown, while in the drier areas the crops included sweet potato, cassava and mixed vegetables. The economic value of the cultivation was estimated as the added value derived from growing crops on the swamp over their dry-land cultivation. The economic values of the papyrus harvest and brick making were estimated by valuing the income from their use for mat-making and other papyrus products, and the sale of bricks made from clay taken from the drier areas during the dry season. The final provisioning service valued was fish farming, and its economic value was calculated based on surveys of the average annual returns to fish farmers in the wetland. In the case of the waste water treatment and purification the wetland's role was to purify the water flowing through the swamp and exiting into Lake Victoria. Two valuation methods for this service were used; In the first the replacement cost method was used, where the cost of artificially replacing the service was estimated. This would involve investing in extending the existing water treatment plant. In the second method, the mitigative expenditure was estimated as the cost required to offset the potential effects of poor water quality. The economic values derived for these analyses in 2002 were as follows:

- crop cultivation – US$60,000 per year;
- papyrus harvesting – US$10,000 per year;
- brick making – US$17,000 per year;
- fish farming – US$3,000 per year; and
- water treatment and purification – US$700,000 to US$1.3 million per year.

(Schuyt, 2005)

The results highlighted the relative overwhelming economic importance of Nakivubo urban swamp in water treatment and purification and the need to sustain that service by retaining an actively growing papyrus swamp. However, one of the difficulties of incorporating this information into a decision-making process is that the benefits of the individual ecosystem services are accrued by different communities. The relatively low value benefits of crop cultivation and papyrus harvesting for the local population are in contrast to the much larger financial benefits to the indirect users who are dispersed much more widely across the catchment area of the river supplying the wetland and beyond.

In another study, Namaalwa et al. (2013) carried out a qualitative assessment of the ecosystem functions and services provided by the Namatala wetland located near the town of Mbale, in the Eastern Region of Uganda. The Namatala wetland is approximately 260 km^2 in area and is an example of a wetland under severe pressure from agricultural development, domestic and industrial pollution, and the harvesting of vegetation. The wetland is divided into two distinct zones; the upper wetland which had lost most of its natural papyrus vegetation and was almost completely converted to agriculture and the lower zone which was less degraded. In some parts of the lower zone the original papyrus vegetation was still intact and fishing and vegetation harvesting were still practiced. The analysis used data on climate, hydrology, water quality, population and land use which was combined with stake-holder workshops. The assessment demonstrated that the communities around Namatala wetland strongly depend on the wetland ecosystem for provisioning services but there is a spatial gradient in services where the upper converted part of the wetland is more important for provisioning services while the lower, more intact wetland, supports regulating and habitat ecosystem services. They found different perceptions among stakeholders about the priority issues in wetland management. Local users were concerned about water and land use conflicts while government agencies were more concerned about agricultural encroachment and biodiversity loss. Namaalwa et al. (2013) concluded that improvements in the management of the Namatala wetland will require (i) improved policy coordination at national level between the various ministries dealing with wetland issues and better coordination between national and local government agencies, (ii) agreement on rights and responsibilities of local resource users, (iii) application of advanced methods of sustainable agriculture and integrated water and nutrient management, and (iv) continued monitoring of the Namatala wetland ecosystem and research into the trade-offs between provisioning and regulating ecosystem services. They suggest, however, that perhaps the biggest issue is the uncertain situation with respect to wetland ownership in Uganda. Many of the rural population believe that they hold land in customary ownership and this causes conflict in the regulation of wetland use and leads to disagreements between local land users and government agencies, particularly over the relative importance of issues such as the encroachment for agricultural use and biodiversity loss as a consequence. Maclean et al. (2011) argue that more effective wetland management could best achieved by preventing individual ownership and promoting common property management regimes.

An alternative approach to the qualitative assessment of ecosystem services was adopted by Morrison et al. (2013) in an effort to prioritise objectives for the rehabilitation of papyrus wetland on Lake Naivasha, a freshwater lake on the rift valley floor in Kenya that had experienced significant environmental degeneration. In this analysis the focus was on the 'stakeholders' that are the riparian communities whose livelihood depends, to varying degrees, on the direct and indirect benefits arising from wetland processes and functions. The primary aim was to provide suggestions for the sustainable management of the wetland's resources that maintains the community use without compromising the critical function of these wetlands in ensuring clean water supply and local climate regulation. The stakeholders answered a questionnaire to identify the perceived benefits of a rehabilitated wetland. These were then allocated to the recognised ecosystem services categories of provisioning, regulating and cultural support, and organised in a 'table of known services'. For the Naivasha wetland the largest proportion (42%) perceived benefits related to provisioning services, and 25% corresponded to cultural services. The remainder (33%) were evenly split between regulating and supporting services. When the same questionnaire was used in another papyrus swamp at a different location (Lake Victoria), the outcome was rather different, with 50.9% of the table of known services allocated to provisioning services, significantly less (13.8%) allocated to cultural services while regulating and supporting services were evenly split again These were interesting observations that illustrated how the differences in perception varied between geographical locations and highlighting the complexity of decision making in either balancing or prioritising the benefits across location.

On a regional scale, Langan et al. (2018) have reviewed the challenges of making assessments of ecosystem services of wetland ecosystems across East Africa. They reviewed a range of environmental modelling approaches that have been used elsewhere to make ecosystem service assessments that support wetland management but, in this case, with a particular focus on the challenges of rural poverty alleviation and climate change impacts. The ecosystem services considered to be central to these challenges were food production, climate impacts and water regulation. The aim was to combine the information on ecosystem services with remote sensing and spatial datasets to create digital maps of wetland properties across the region. They identified 21 ecological processes which contributed to the ecosystem services and, not surprisingly, they found that no one model was able to capture these processes effectively in the case of tropical wetlands. They concluded that there was a lack of information on the properties of East African wetlands sufficient to quantify ecosystem services which would assess the impact of human activity and could allow them inform management decisions. Furthermore, even where data are available the values were based on localised ratings and scoring systems that are frequently not widely applicable at scale. They suggested that more, high quality, spatial data on the hydrological, soil and vegetation properties of wetland ecosystems are required to support evidence-based wetland ecosystem service management tools at the region scale (Langan et al., 2018).

Subsequently, Langan et al. (2019) have attempted to overcome the problems of absence of quantitative data by using qualitative assessments which were developed into a much simplified wetland ecosystem tool specific to papyrus swamps. The 'wetland environmental services tool' includes a field survey methodology for measuring wetland properties under different land covers, and a simple accounting procedure for assessing wetland functions and ecosystem services sensitive to underlying wetland properties. The field surveys provided standardised data which was combined with existing literature values to estimate crop, water and climate related ecosystem services provided by the wetland. When applied to a wetland complex in the Kabale district in south western Uganda results demonstrated how anthropogenic activities have had a major influence on the wetland properties. The assessment was based on estimates of ecosystem services provided by six land cover categories of cultivated land, forest, grassland, reeds, papyrus and open water. In this wetland complex as much as 40% of the wetland has undergone change to potato cultivation. At the same time the wetland is both a large stock of water that releases moisture to the local atmosphere, and is a large stock of carbon which is currently a net sink sequestering an estimated 3,000 t carbon annually. The wetland also provides water quality and flood storage services. Figure 6.1 is a radar diagram showing the provision of ecosystem services for each of the land cover categories (Langan et al., 2019). It is suggested that although this assessment of ecosystem services is limited it provides a useful methodology for establishing a baseline for monitoring changes in ecosystem properties in the future (Langan et al., 2019).

Ecosystem Services for Sustainable Management of Papyrus Wetlands

Sustainable management has been described as 'the long-term use of one or several of the wetland ecosystem services without degrading or losing other essential ecosystem services' (van Dam et al., 2011; van Dam et al., 2014). In other words, for ensuring sustainable management there is a need to identify suitable strategies that allow local communities to support their livelihoods while ensuring the continued integrity of the ecosystem. The process of development of a sustainably managed wetland depends on (i) describing the ecological characteristics of the wetland system, (ii) identifying the wetland stakeholder characteristics and their relationships, (iii) assessing the ecosystem services and functions, (iv) analysing the key drivers

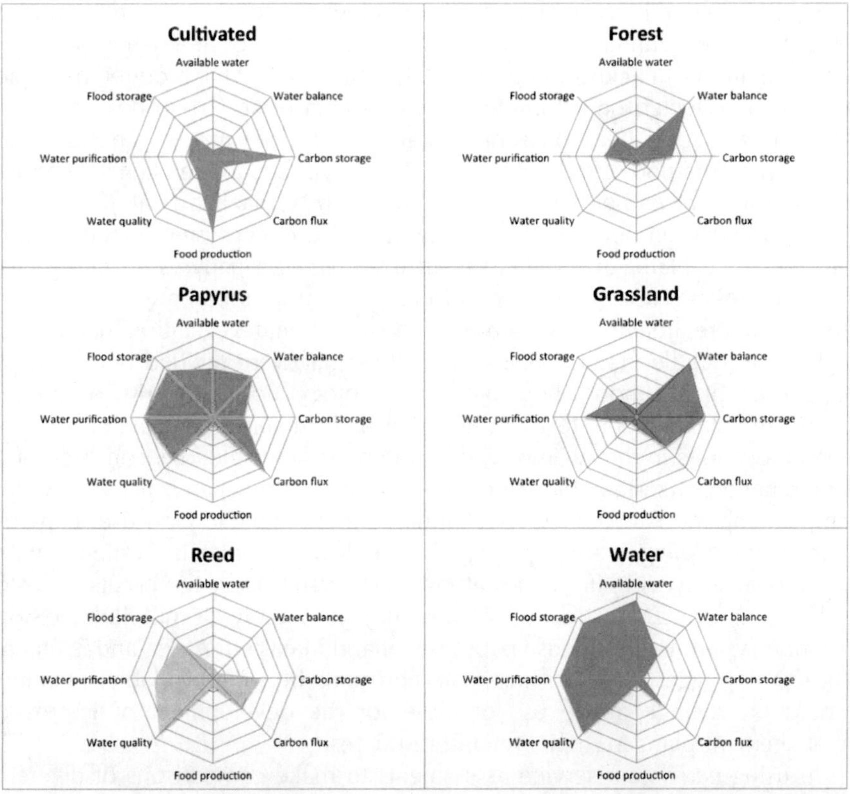

Figure 6.1 A radar diagram displaying the provision of ecosystem services by different forms of land cover in a wetland complex in southwest Uganda. Service provision values are centred and rescaled so that the outside edge of the diagram represents the highest provision of an ecosystem service by any land cover class and the centre represents the lowest ecosystem service provision of any land cover.

Source: Langan et al. (2019)

and pressures, and their impact on the ecosystem services, and (v) identifying potential policy and management responses that can provide a balanced and sustainable exploitation of ecosystem services in the wetland (Dixon and Wood, 2003; Namaalwa et al., 2013). Decision-making in wetlands management therefore involves the evaluation of the trade-offs among ecosystem services. For this to be effective it ideally requires some form of quantification of all ecosystem services and agreement among stakeholders about the desirability of maximising the impact of favoured ecosystem services. In the case of papyrus wetlands this is fraught with difficulties because, as we have seen, not all ecosystem services are easily quantifiable and furthermore the successful application of an integrated approach requires accessibility to large amounts of hydrological, geomorphological, ecological and economic data to make quantitative, model-based evaluations of alternative management solutions. In acknowledging the fact that much of the required data had not been collected on a long term basis, Zsuffa et al. (2014) described the development of a decision support framework (DSF) that was designed to cope with data-poor conditions. The DSF was applied to the Namatala multi-use wetland in northeastern Uganda, on which Namaalwa et al. (2013) had previously carried out the qualitative assessment of ecosystem functions and services. The Namatala wetland is an example of a papyrus swamp under intense pressure from wastewater discharge from an urban area, conversion to agriculture (mainly rice farming) and traditional vegetation harvesting, fishing and cattle grazing. Using indicators that were divided between the categories of livelihood, human health, ecology, costs, and risk of failure, they used stakeholders to evaluate three main management option for different locations in the swamp. The best-compromise solutions consisted of a combination of sustainable agriculture in the upper wetland, papyrus buffer strips along the Namatala river channel, and sustainable land use (papyrus harvesting, fishing, cattle grazing) in the lower Namatala wetland, with additional papyrus buffer zones at the waste-water discharge points close to urban developments. This was the first attempt to apply a simplified decision support system to a multi-use papyrus wetland in eastern Africa and Zsuffa et al. (2014) suggest that, as more data and tools become available, planners could increasingly make use of these for the development of improved management plans for their wetland resources.

In using ecosystem service assessments to make comparisons of different management strategies for papyrus swamps, one of the fundamental questions has been is 'what are the relative merits of managing by conserving, using or draining papyrus swamps?' (Maclean et al., 2011). The draining of papyrus vegetation and the conversion of former natural wetlands to crop production enables smallholder farmers to produce as many of three crops per year on moist soils built up from sediments that are likely to initially contain high levels of essential nutrients and are therefore very fertile, although this declines with intensive utilisation. More importantly though, this expansion of food crop cultivation onto wetland areas occurs at the expense of many of the provisioning and regulating ecosystem services formerly provided by the swamps. Most significantly, the drainage and cultivation of wetlands has serious impacts on the regulatory capacity of storing water during the wet season and maintaining a base flow during drier periods of the year. These can have serious implications for local communities as well as those downstream who are dependent on this regular flow of water. Clearly, the conversion of the wetlands to agricultural production is driven by the requirements of the local householders for food crops to feed rapidly increasing populations and for them wetland drainage and cultivation has been widely regarded, particularly among policy makers, as a panacea for addressing the food needs of land-scarce communities.

However, there is clearly the need to consider the negative impacts on the wider population who are affected by the role and functioning of wetlands in the regional hydrological cycle (Dixon and Wood, 2003). Therefore, although wetland drainage and cultivation can make an important contribution to food and livelihood security in the short term, in the long term there are concerns about the sustainability of this utilisation of wetland benefits.

The identification in functioning wetlands of the large number of potentially available ecosystem services and the growing awareness of their benefits has encouraged a shift towards thinking about multifunctionality of individual wetlands and the benefits of the sustainable delivery of multiple ecosystem services by using the papyrus resource (Dixon and Wood, 2003). This has led to increased recognition that sustainable management has the potential to derive maximum value from low-intensity multifunctional wetland uses which far exceeds the value derived exclusively from swamp drainage and use for agricultural crop production. For example, Maclean et al. (2011) summarised the outputs from three studies on papyrus swamps in Uganda and Kenya which calculated the monetary value of their current ecosystem services and showed that for all three studies the economic value of low-intensity wetland management use would greatly exceed the value of wetland drainage and conversion to crop production. Low-intensity uses included harvesting papyrus, subsistence fishing, grazing and hunting. The study demonstrated that the value of low-intensity use was maximised when 27–33% of a swamp is utilised for harvesting, but less than 2% reclaimed for subsistence cultivation. The work shows that optimising wetland use becomes a series of compromises between getting different services from the ecosystem while avoiding the negative consequences of trade-offs which can occur between the services, both provisioning and regulating (Swallow et al., 2009, Langan et al., 2018). In order to achieve this, McCartney and Houghton-Carr (2009) proposed the use of an index to guide the sustainable development of wetlands which support the effective use of a wetland for agricultural purposes while maintaining the essential elements of its ecological character. They assessed the so-called 'working wetland potential' (WWP) using a five-step process, consisting of (i) assessment of the current ecological character of the wetland, (ii) assessment of the wetlands' contribution to social welfare, (iii) assessment of the suitability of the wetland for agriculture, (iv) assessment of the possible hazards associated with the wetland agriculture and (v) an evaluation of the WWP. The WWP method was proposed as a way of explicitly integrating social and biophysical aspects of wetland systems in a single index to determine the potential for using a wetland for specified agricultural activities.

In conclusion it is clear that the key challenge to successfully applying the ecosystem services concept in wetland management and conservation is in providing scientifically rigorous information about the mechanisms driving wetland dynamics and developing, preferably quantitative, assessments of these processes (Xu et al., 2020). The requirement is to appropriately and transparently incorporate the range of potential ecological functions which operate at different spatial and temporal scales, while avoiding overloading decision makers with too much detail. As we have seen, the evidence-base for most ecosystem service assessments have largely been built upon a wide range of individual studies that mostly focus on small scale local subcomponents of the ecosystem. In order to apply the analysis of ecosystem services at the regional or national scale requires the applications of mechanistic models with inputs from the small scale studies (Langan et al., 2018).

Causes of Wetland Loss – Management Failures

The environmental degradation of actively growing papyrus swamps, particularly in areas of high population density, have most frequently been attributed to excessive unchecked exploitation by local communities with open access to the wetland. Because of the absence of local management environmental degradation is the consequence of prolonged periods of opportunistic and uncontrolled harvesting of papyrus and the eventual conversion of severely degraded swamps to cultivated areas by encroachment, where water levels allow, or drainage by excavating channels in the sediment. Apart from the excessive harvesting of papyrus culms, one of the more common reasons for the long term decline has been when cattle herders continuously harvest young papyrus umbels as the most nutritious part of the plant to use as livestock feed, particularly during the dry season when grass growth is poor (Muthuri and KInyamario, 1989). Additionally, an increasingly common cause of papyrus wetland degeneration is attributed to an increasing reliance on papyrus as a source of fuel, either on a daily basis or when wood charcoal and fuelwood are unobtainable or too expensive. Furthermore, when water levels decline, exposing the rhizomes of the stranded papyrus, they are often harvested and used preferentially as a source of fuel because they produce more heat than do the culms when burnt (Morrison et al., 2012).

A sociological reason for wetland loss is ascribed to a lack of consistency in the application government policies. These failures are, to some extent, due to an insufficient understanding by decision makers of the functions and values of wetlands and the consequences of their loss in terms of the ecosystem services they potentially provide (Turner et al., 2000; Schuyt, 2005). For example, Were et al. (2021) found that in recent years one of the leading reasons for pristine papyrus wetland removal in Uganda has been the approval of plans to allow large scale planting of rice, which is easily cultivated and has relatively high economic returns. However, an unintended consequence of this action is a significant reduction in carbon stored in peat below the former wetland vegetation. This occurs as a result of release of CO_2 when under cultivation peat decomposes on exposure to an aerobic environment (see Chapter 4). Unfortunately, the value of the loss of this service in sequestering CO_2 and mitigating climate warming is difficult to calculate and is therefore most likely to be ignored in any assessment of ecosystem services.

Developing Policies for Sustainable Management

Of the states within the Nile Basin, Uganda appears to have been the most active in developing management programmes for its extensive wetlands. This effectively started in 1993 with the establishment of the National Wetland Programme, which was a mechanism to formulate a national policy for the conservation and management of Uganda's wetlands (Hartter and Ryan, 2010). In 1995, a new constitution asserted the significance of environmental conservation in stating that natural resources are held in trust by the government for the people of Uganda. A national policy of decentralisation devolved rights and responsibilities to overseeing this to local government. The lead technical agency responsible for managing wetlands is the Directorate of Environmental Affairs (DEA) operating under the Ministry of Water and Environment (MWE) and the implementation of the decentralisation policy to local governments is overseen by the Ministry of Local Government. The Wetland Management Department (WMD) of the DEA is mandated to regulate and control wetland use activities. A number of additional institutions including the Ministries of Agriculture, Animal Industry and Fisheries, Finance, Planning and Economic Development as well as local governments from District to Village level, are also involved in this complex structure. Unfortunately the linkages between these institutions have been weak and as a result there are diverging perceptions among stakeholders about the priority issues in wetland management, while the WMD have experienced difficulties in connecting national level natural resource management goals with local needs. Part of the problem appears to be linked to misunderstandings about the ownership of the wetlands. At the national level the WMD are clear that, in line with the constitution, permanently inundated wetlands cannot be privately held, but are instead held by the government and managed by local councils. These areas should then be managed to allow the community access to the wetland. However, surveys of local communities have indicated that most respondents thought that either they themselves or another community member were owners of the wetland area that they used and that they had the right to control access. Hartter and Ryan (2010) suggest that this is symptomatic of a continuing struggle to connect the national level natural resource management goals with the local needs. While the national policies and legislations have provided a broad framework for managing wetlands, they remain ineffective at the local level. For example, Namaalwa et al. (2013) cite an example of a framework management plan for the Namatala wetland, referred to above, which was developed through a participatory process involving stakeholders from districts sharing the wetland and published in 2008. The plan included strategies for the protection of the social, economic and environmental benefits of the wetlands, but it was never implemented. The reason for this appears to be a combination of issues including, at the local level, limited spending budgets and lack of manpower available to district environmental officers, and at the national level, the difficulty that the top-down approach to implementing state law frequently results in poor compliance and an absence of local community participation in implementing the recommendations (Willbroad and Kiyawa, 2019).

Proposals for improved management implementation have included better horizontal policy coordination at national level between various ministries dealing with the wetland issues as well as improved vertical coordination between national and local government levels. Effective implementation of the national policy framework requires strong linkages and coordination with local government institutions, including adequate funding, capacity building and participation of end users. In addition, the rights and responsibilities of the local resource user needs to be defined and agreed so that they are able to participate in a meaningful way in the decision making process (Turner et al., 2000; Hartter and Ryan, 2010).

Going forward the development of a sound base for wetland management of wetlands in the Nile Basin depends on formulating a management strategy and then identifying the essential elements that need to be incorporated into the management plans. This process needs to ensure that local communities are involved in all conservation and management decisions and activities by training personnel in wetland management and dissemination information on the importance of wetlands. As the wetlands occur in dispersed rural areas, there needs to be extension services which are put in place to create awareness of the local needs for conservation and management. The first action of the extension services should be to identify the natural, scientific, aesthetic and educational values and benefits of the local wetlands. It will then be possible to identify critical areas that are rich in biodiversity and declare these as critical and protected sanctuaries. Wetlands that can be used for the supply of ecosystem services should be managed as multiple use resources. To achieve this it may be necessary for the users to agree to intensively use parts of the wetland on a rotational basis while maintaining other areas in an almost pristine state.

Conclusion

It was the Millennium Ecosystem Assessment that highlighted the range of provisioning, regulating, supporting and cultural ecosystem services that could be obtained from wetlands. The concept of ecosystem services, defined as the contribution of ecosystems to human wellbeing, has since 2005 drawn increasing attention from researchers due to its significance for the practical management of wetland ecosystems and the potential trade-offs between alternative management options. The extensive wetlands associated with the Lake Victoria basin have been the provider of substantial supplies of provisioning ecosystem services in the form of food crops that has supported the demands of the rapidly expanding rural population. The demand for land for food production has led to not only widespread encroachment onto cultivatable margins of wetlands but also to the excavation of drainage channels to increase water flow through the swamp and destroy the wetland vegetation to expose land for planting. The immediate economic benefits from food crop production are, however, accompanied by the loss of the hydrological and water quality functions of the wetland. Unfortunately, although converting wetlands to cultivation of crops can adversely affect their ability to provide the full range of ecosystem services it is likely that further destruction of wetlands for agriculture will continue as demand for food production is maintained and alternative livelihood opportunities for land users are lacking.

An alternative scenario is to encourage sustainable multiple-use management of wetlands, as it is now clear that the overall benefits are greater when all ecosystem services are balanced so that no single category is developed at the cost of another. The sustainable management of wetland ecosystems requires the use of practices that balance the human needs with the conservation and protection of the ecosystem. Management practices should be participatory in approach to involve local residents and current users of wetland resources to raise the awareness of the importance of wetlands and their multiple benefits. Local communities should be involved in all conservation and management activities, including the formation of extension services to create awareness. The aim should be to develop integrated management authorities to include local communities, government and non-government agencies, conservationists and ecologists.

References

Costanza, R., De Groot, R., Sutton, P., Van der Ploeg, S., Anderson, S.J., Kubiszewski, I., Farber, S. and Turner, R.K. (2014). Changes in the global value of ecosystem services. *Global Environmental Change*, 26, 152–158.

Dixon, A.B. and Wood, A.P. (2003) Wetland cultivation and hydrological management in eastern Africa: Matching community and hydrological needs through sustainable wetland use. *Natural Resources Forum*, 27, 117–129.

Donaldson, L., Woodhead, Wilson, R.J. and Maclean, I.M.D. (2016) Subsistence use of papyrus is compatible with wetland bird conservation. *Biological Conservation*, 201, 414–422.

Emerton, L., Iyango, L., Luwum, P. and Malinga, A. (1998). *The present economic value of Nakivubo urban wetland, Uganda*. Nairobi: IUCN Eastern African Regional Office.

Gaudet, J.J. and Muthuri, F.M. (1981). Nutrient relationships in shallow water in an African Lake, Lake Naivasha. *Oecologia*, 49, 109–118.

Hartter, J. and Ryan, S.J. (2010). Top-down or bottom-up? Decentralisation, natural resource management, usufruct rights in the forests and wetlands of western Uganda. *Land Use Policy*, 27, 815–826.

Hes, E.M.A., Niu, R. and van Dam, A. (2014). A simulation model for nitrogen cycling in natural rooted papyrus wetlands in East Africa. *Wetlands Ecology and Management*, 22, 157–176.

Hes, E.M. and van Dam, A.A. (2019). Modelling nitrogen and phosphorus cycling and retention in *Cyperus papyrus* dominated natural wetlands. *Environmental Modelling & Software*, 122, 104531. doi:10.1016/j.envsoft.2019.104531.

Junk, W.J., An, S., Finlayson, C.M., Gopal, B., Květ, J., Mitchell, S.A., Mitsch, W.J. and Robarts, R.D. (2013). Current state of knowledge regarding the world's wetlands and their future under global climate change: a synthesis. *Aquatic Sciences*, 75, 151–167.

Kansiime, F., Saunders, M.J. and Loiselle, S.A. (2007). Functioning and dynamics of wetland vegetation of Lake Victoria: an overview. *Wetlands Ecology and Management*, 15, 443–451.

Kiwango, Y.A. and Wolanski, E. (2008). Papyrus wetlands, nutrient balance, fisheries collapse, food security, and Lake Victoria level decline in 2000–2006. *Wetland Ecology & Management*, 16, 89–96.

Langan, C., Farmer, J., Rivington, M. and Smith, J.U. (2018). Tropical wetland ecosystem service assessments in East Africa; A review of approaches and challenges. *Environmental Modelling & Software*, 102, 260–273.

Langan, C., Farmer, J., Rivington, M., Novo, P. and Smith, J.U. (2019). A wetland ecosystem service assessment tool: development and application in a tropical peatland in Uganda. *Ecological Indicators*, 103, 434–445.

McCartney, M.P. and Houghton-Carr, H.A. (2009) Working wetland potential: An index to guide the sustainable development of African wetlands. *Natural Resources Forum*, 33(2), 99–110.

Mace, G.M., Norris, K. and Fitter, A.H. (2012). Biodiversity and ecosystem services: a multilayered relationship. *Trends in Ecology & Evolution*, 27, 19–26.

Maclean, I.M.D., Hassall, M., Boar, R.R. and Lake, I.R. (2006). Effects of disturbance and habitat loss on papyrus-dwelling passerines. *Biological Conservation*, 131, 348–358.

Maclean, I.M.D., Boar, R.R. and Lugo, C. (2011). A review of the relative merits of conserving, using, or draining papyrus swamps. *Environmental Management*, 47, 218–229.

MEA. (2005). *Ecosystems and human well-being: A framework for assessment*. Washington, DC: Island Press.

Mitchell, S.A. (2013). The status of wetlands, threats and the predicted effect of global climate change: the situation in Sub-Saharan Africa. *Aquatic Sciences*, 75, 95–112.

Morrison, E.H. J., Upton, C., Odhiambo-K'oyooh, K, and Harper, D.M. (2012). Managing the natural capital of papyrus within the riparian zones of Lake Victoria, Kenya. *Hydrobiologia*, 692, 5–17.

Morrison, E.H. J., Upton, C., Pacini, N., Odhiambo-K'oyooh, K., and Harper, D.M. (2013). Public perceptions of papyrus: community appraisal of wetland ecosystem services at Lake Naivasha, Kenya. *Ecology and Hydrobiology*, 13, 135–147.

Muthuri, F.M. and Kinyamario, J.I. (1989). Nutritive value of papyrus (*Cyperus papyrus*, Cyperaceae), a tropical emergent macrophyte. *Economic Botany*, 43, 23–30.

Namaalwa, S., van Dam, A.A., Funk, A., Ajie, G.S. and Kaggwa, R.C. (2013). A characterization of the drivers, pressures, ecosystem functions and services of Mamatala wetland, Uganda. *Environmental Science and Policy*, 34, 44–57.

Pacini, N., Hesslerová, P., Pokorný, J., Mwinami, T., Morrison, E.H., Cook, A.A., Zhang, S. and Harper, D.M. (2018). Papyrus as an ecohydrological tool for restoring ecosystem services in Afrotropical wetlands. *Ecohydrology & Hydrobiology*, 18(2), 142–154.

Rebelo, L.M., McCartney, M.P. and Finlayson, C.M. (2010). Wetlands of Sub-Saharan Africa: distribution and contribution of agriculture to livelihoods. *Wetlands Ecology and Management*, 18, 557–572.

Rongoei, P.J.K., Kipkemboi, J., Okeyo-Owuor, J.B. and Van Dam, A.A. (2013). Ecosystem services and drivers of change in Nyando floodplain wetland, Kenya. *African Journal of Environmental Science and Technology*, 7, 274–291.

Saunders, M.J., Jones, M.B. and Kansiime, F. (2007). Carbon and water cycles in tropical papyrus wetlands. *Wetlands Ecology and Management*, 15, 489–498.

Saunders, M.J., Kansiime, F. and Jones, M.B. (2014). Reviewing the carbon cycle dynamics and carbon sequestration potential of *Cyperus papyrus L.* wetlands in tropical Africa. *Wetlands Ecology and Management*, 22, 143–155.

Schuyt, K.D. (2005). Economic consequences of wetland degradation for local populations in Africa. *Ecological Economics*, 53, 177–190.

Swallow, B.M., Sang, J.K., Nyabenge, M., Bundotich, D.K., Duraiappah, A.K. and Yatich, T.B. (2009). Tradeoffs, synergies and traps among ecosystem services in the Lake Victoria basin of East Africa. *Environmental Science & Policy*, 12, 504–519.

Thorslund, J., *et al.* (2017). Wetlands as large-scale nature-based solutions: Status and challenges for research, engineering and management. *Ecological Engineering*, 108, 489–497.

Turner, R.K., Van Den Bergh, J. C., Söderqvist, T., Barendregt, A., Van Der Straaten, J., Maltby, E. and Van Ierland, E. C. (2000). Ecological-economic analysis of wetlands: scientific integration for management and policy. *Ecological Economics*, 35, 7–23.

van Dam, A., Kipkemboi, J., Zaal, F. and Okeyo-Owuor, J.B. (2011). The ecology of livelihoods in East African papyrus wetlands (ECOLIVE). *Reviews in Environmental Science and Bio/Technology*, 10, 291–300.

van Dam, A.A., Kipkemboi, J., Mazvimavi, D. and Irvine, K. (2014). A synthesis of past, current and future research for protection and management of papyrus (*Cyperus papyrus* L.) wetland in Africa. *Wetland Ecology and Management*, 22, 99–114.

van Dam, A.A. and Kipkemboi, J., 2016. Sustainable use of papyrus from Lake Victoria, Kenya. In *The wetland book* (pp. 1–12). Dordrecht: Springer.

Were, D., Kansiime, F., Fetahi, T. and Hein, T. (2021). Carbon dioxide and methane fluxes from a tropical freshwater wetland under natural and rice paddy conditions: implications for climate change mitigation. *Wetlands*, 41, 52. doi:10.1007/s13157-021-01451-4.

Willbroad, B. and Kiyawa, S.A. (2019). Sustainable management and conservation of wetland resources in Uganda: a review. *Journal of Environment and Health Science*, 5, 47–51.

Xu, X., Jiang, B., Tan, Y., Costanza, R. and Yang, G. (2018). Lake-wetland ecosystem services modeling and valuation: Progress, gaps and future directions. *Ecosystem Services*, 33, 19–28.

Xu, X., Chen, M., Yang, G., Jiang, B. and Zhang, J. (2020). Wetland ecosystem services research: A critical review. *Global Ecology and Conservation*, 22, e01027.

Zsuffa, I., van Dam, A.A., Kaggwa, R.C., Namaalwa, S., Mahieu, M., Cools, J. and Johnston, R. (2014). Towards decision support-based integrated management planning of papyrus wetlands: a case study from Uganda. *Wetland Ecology and Management*, 22, 199–213.

7 The Future for Papyrus Wetlands

Introduction

This concluding chapter is a review of the future prospects for papyrus and, in particular, the impacts of the growing human population and the changing climate on the papyrus wetlands with a focus on those of the Nile basin. We ask whether there are prospects for protecting pristine papyrus wetlands from long term and permanent decline or can they be retained by sustained utilisation? Are there prospects of, for example, combining bioenergy and nature conservation. Alternatively, will our improved knowledge of wetland biomass production, and sustainable management options, enable the local populations promote the development of bio-based products to diversify and enhance a local bio-economy. On the other hand, will the human degradation of wetlands, led by pressures from population growth and climate change, threaten the destruction of papyrus wetlands and what are the consequences? Is the decline seen in recent decades the beginning of a downward spiral or are there hopes of sustaining the wetlands and possibly re-establishing them, in the sure knowledge that they have the potential to provide a viable contribution to the future of the more than 250 million people who are reliant on the Nile waters? The message is that an earlier civilisation was able to exploit a natural resource sustainably for close on three thousand years to support its survival and development. Can that same resource now offer support for the future, if conserved and managed sustainably? Alternatively, will papyrus wetlands disappear just as they did from Egypt centuries ago, when wetlands were extensively drained and used for valuable crop production?

The debate over the future of wetlands can fluctuate between, on the one hand, the view that seeks to see complete protection of wetlands as fragile ecosystems that provide valuable and irreplaceable ecosystem services to the local and regional community and, on the other hand, the complete transformation for agricultural utilisation. The view of early researchers on papyrus wetlands, whose focus was on the economic benefits derived from wetlands in the form of fishing, forestry and agriculture for Colonial governments, suggested that 'at the present time the swamps are of very little economic value' (Carter, 1955).

DOI: 10.4324/9781003131540-7

However, this was cautioned by the statement that 'before any extended use of the swamps is planned it is clear that accurate knowledge of the nature of the swamp is required.' Research since then in a range of disciplines including ecology, biogeochemistry, hydrology and socio-economics has led to the increasingly world-wide recognition of the significance of wetlands in delivering ecosystem services that when intact have economic values greater than their value when converted to agriculture.

Currently, wetlands support millions of people in Africa as they offer sources of freshwater, fisheries, irrigation of agricultural land, construction materials, medicinal uses and wild plants for food. They also help to control flood water, recharge ground-water and purify drinking water. The runoff of vital plant nutrients decreases in a catchment as the area of the wetland increases and the wetlands play a key role in mitigating rapid responses to flooding by ensuring the longer residence time of water in a catchment. However, at the same time wetlands are arguably the most threatened ecosystems in Africa. Probably, the most impacted wetlands being those around Lake Victoria, which have experienced dramatic changes due to rapid fluctuations in water levels, the introduction of the Nile Perch fish, sewage inflows and increased sedimentation and nutrient flow in the lake from massively increased agricultural activity in the Nile Basin. Fundamentally, the three main contributors to the ecological decline in these wetlands have been growing population, economic development and climate change (Turyahabwe et al., 2013). Increased crop production in the basin wetlands has been due to an increased demand for staple food crops. These crops include maize and paddy rice in the warmer areas at lower altitudes and potatoes that have been important in the highland cooler areas. Increased production has been achieved by draining and clearing papyrus wetlands for fertile land, which has negative consequences for the hydrology of the area. For example, many of the upland valley swamps in the Kigezi Highlands of southwestern Uganda have suffered from increased drainage carried out for the cultivation of potatoes. When drained the water is conveyed rapidly downstream reducing the wetland's ability to buffer peak flows and retain nutrients. The role which these extensive wetlands formerly played in regulating the flow of water into Lake Victoria has now been all but lost (Farmer et al., 2022).

While knowledge of ecosystem services provided by papyrus wetlands at local scales is extensive, it is widely recognised that this knowledge has not been effectively translated to management at the regional scale. There are a number of requirements that need to be achieved before this can happen. Probably the most important are the development of national wetland inventories. These require better and more detailed assessments of the aerial extent and ecosystem services provided by the wetlands, which needs to be supplemented with assessments of biodiversity, and ecosystem health. Furthermore, research is needed to quantify the processes that support the climate change mitigation effects of wetlands at the regional scale. In the meantime, some countries have taken steps to try to protect wetlands from continuing damage and rapid decline. One of the first to recognise the issue was the Ugandan government when it implemented several policies that emphasise optimisation of sustainable benefits of wetlands, while conserving the environment and biodiversity. These were developed under the National Policy for the conservation and management of wetlands in 1995, and the policies emphasised protection of the wetlands and forbad any form of wetland reclamation. However, despite the presence of this legislation large areas of wetlands have subsequently been reclaimed, mainly for urban agriculture, emphasising the difficulty of implementing legislation effectively.

Looking forward there are a number of management issues which will influence the distribution and functioning of papyrus wetlands. Their interplay will ultimately determine whether or not papyrus swamps will play a significant role in supporting the wetland environments. These issues play out at different spatial scales from the local to the regional. Most attempts to date have involved management at the local scale and they all depend on the concept of multiple wetlands use as a means of achieving environmentally sustainable and equitable wetland use. Where multiple wetland uses exist a range of benefits can be sustained with little evidence of environmental degradation at the local scale but the key question is how to encourage and integrate such multiple regimes at larger scales. There is a need to raise awareness of the importance of wetlands and their multiple benefits and to target policy-makers in the agriculture and water resource sectors and persuade them to consider the value of the benefits in a whole catchment wide context. In this chapter we consider how some of the regional scale issues may influence the survival of papyrus wetlands and what contribution they could make to the wetland landscape of the future.

Future Regional Cooperation for Wetland Management

In order to protect and effectively manage the Nile basin wetlands, managers need to know details of the composition of the wetland ecosystems, their area and exact geographical location. In other words, there needs to be a comprehensive wetland inventory. At the regional scale this needs to be composed of combined national wetland inventories. The collection of field observations mainly involves the establishment of monitoring points, monitoring stations, and monitoring centres, the construction of monitoring databases and wetland information management platforms (Xu et al., 2019). Field observation is an important means to reveal the changes in structure and function of wetland ecosystems. However, currently there are difficulties in achieving these aims at the regional scale as it is estimated that less than 35% of African nations have a national wetland inventory and many of these are not comprehensive. Collecting all this information is a very labour-intensive operation, particularly because many of the wetlands are difficult to access on the ground. However, it is becoming increasingly clear that the most effective solution will be to use more earth observation data that is becoming available from satellite-based sensors. With this information policy-makers can support communities so that they can then use local knowledge to manage the wetlands themselves for wetland protection and management.

Wetland management policies reviewed in Chapter 6 have, broadly speaking, focussed on protection and rational utilisation at the local scale. At the regional and international scale a major influence has been the Ramsar Convention, signed in 1971. This intergovernmental treaty was first aimed at the conservation of wetlands as a habitat for water birds and is named after the Iranian city in which the treaty was signed. The convention subsequently developed to cover all aspects of wetland conservation of biodiversity and the well-being of human communities with an overall objective of conservation and wise use of wetlands by national action and international cooperation as a means to achieving sustainable development. The Convention aims at protecting wetland ecosystem functions, maintaining wetland cultures and realising sustainable socio-economic development through local, regional, and national actions as well as international cooperation. The Ramsar Convention Secretariat, based in Gland, Switzerland, is supported by the International Union for Conservation of Nature (IUCN). By designating Ramsar Sites the Convention recognises and protects wetlands of international importance, promoting their conservation and sustainable use. The Ramsar list of designated sites has become the world's largest network of protected areas. There are now over 2,300 Ramsar wetland sites worldwide in 169 countries and covering about 230 million hectares, which is approximately 20% of the global wetland area. The Ramsar sites are distributed unevenly in different continents, particularly in Africa which has 397 sites with an area of 110 million hectares, almost half of the total global area. In a global survey by Xu et al. (2019) of the Ramsar sites more than half of them were judged to be affected by three of four 'impact factors'. The most significant impact factors were pollution (54%), then biological resource use (53%), natural system modification (53%), and agricultural encroachment (42%). In a survey of 30 Ramsar sites in Africa it was found that 17% of the river wetlands and 20% of the flooded wetlands had degraded into 'no-wetlands' and it was concluded that there is therefore a need for local, regional and national cooperation to work together continually to strengthen global wetland conservation and restoration (Xu et al., 2019). The development of management plans for the optimal sustainability of Nile Basin wetlands depends on establishing long term monitoring while maintaining the continuity of effective communications within the sites, organisations and stakeholders, demonstrating the effectiveness of management and ensuring compliance with local, national and international policies. In support of these objectives the Ramsar Secretariat in conjunction with the European Space Agency launched in 2003 the 'GlobWetland' project in order to demonstrate the current capabilities of remote sensing technology to support inventorying, monitoring, and assessment of wetland ecosystems. In the last few decades remote sensing technology has proven to be a powerful tool to monitor and assess the Earth surface, the oceans and the atmosphere on a regular basis. Remote sensing satellites with increasing capabilities in terms of spatial, temporal and spectral resolution, allow a more efficient, reliable and affordable monitoring of the environment over time at global, regional and local scales. The ultimate aim of the GlobWetland project was to provide wetland managers and national authorities with an efficient support tool for management and conservation authorities. This is despite the fact that the particular nature of wetlands, which can include a combination of shallow waters, inundated vegetation and wet soils, renders their identification and delineation over large areas at high resolution from remote sensing technology a complex technical problem. Therefore, within the context of the GlobWetland project, an alternative approach was developed that provides the national authorities with a map of potential wetland sites as a support tool to facilitate the preparation of national inventories at large scale. The project benefits were most significant for Sub-Saharan countries where in many cases the GlobWetland data sets were the first and only updated digital geo-referenced maps available in the area of interest. As a result remote sensing proved to be an effective technology for monitoring spatial and temporal distributions and dynamic changes of wetlands on a large scale. However, the significant consultation process with wetland managers, national authorities and scientists carried out during the lifetime of the project identified a significant gap in communication between the remote sensing and wetland communities, which need to be improved in the future while the remote sensing technology continues to be developed to increase the resolution of the monitoring data and the speed of data updating (Jones et al., 2009).

At the sub-regional and local scale, research on wetlands in Africa has often been undertaken by or with the support of NGOs and various donor agencies. Research has focussed importantly on the role of smaller wetland complexes in the livelihood of the local populations. Of the East African countries, Uganda leads the way in incorporating this knowledge into policy and attempts at sustainable wetland management, which conserves the wetlands while providing sustainable livelihoods for people living adjacent to the wetlands. Unfortunately, as we have seen, there remains a lack of baseline information on the temporal extent, distribution and characteristics of these wetlands although the situation has improved significantly in the last two decades.

At the regional scale, the Nile Basin Initiative (NBI) has been established as an intergovernmental partnership of 11 Nile Basin countries. For the first time in the basin's history an all-inclusive basin-wide institution was established in 1999 to provide a forum for consultation and coordination among the basin states for 'the sustainable management and development of the shared Nile Basin water and related resources for win-win benefits and to promote regional peace and security'. The shared vision is to achieve sustainable socio-economic development through the equitable utilisation of, and benefits from, the common Nile Basin water resources. The NBI developed a Nile Basin Monitoring Strategy to guide the activities for enhancing the monitoring system in the basin with the intention of providing reliable and continuously observed data that can be used by countries in the basin to improve their water resources planning and management. The Nile Basin Regional HydroMet System was launched in November 2019. A total of 73 hydrological stations have been earmarked, 69 of which are rehabilitations while five are newly installed. The intention is to use the data to improve flood disaster preparedness, coordinate management of water storage dams, and improve adaptation to climate change. It is seen as being critically important for dealing with the challenges of climate change, particularly the recent unprecedented rise of Lake Victoria water level and also the Blue Nile and associated upstream impacts. It is financially supported by the European Union and the German Federal government through German International Cooperation. Climate change adaptation includes; early warning of floods (eastern Nile Basin seasonal forecasting), river flow forecasting for the whole of the Nile Basin, drought monitoring and forecasting early warning potential (www.flooddroughtmonitor.com). Quarterly Basin monitoring bulletins using Satellite Observations have been produced and disseminated. This provides information on the spatial distribution of seasonal patterns of some of the components of the water cycle such as rainfall and actual evapotranspiration in the Nile Basin sub-basins in comparison to historical, long-term averages and seasonal anomalies.

In addition, under the Nile Basin Initiative, a study of peatlands of the Nile Basin region provides a measure of extent, carbon stock, land-use threats and practical strategies for their sustainable use and management. These peatlands are under increasing land use threats and are disappearing at an alarming rate due to agricultural clearing, peat extraction for energy and drainage for monoculture agriculture. This leads to increased carbon emissions at a time when the Nile Basin countries are signatories of the Paris Agreement to reduce carbon emissions. The total area of peatlands and other organic soils in the Nile Basin amounts to 30,445 km^2 and estimated peat carbon stocks are 4.4–10 Gt C, 10% of the total tropical peatland C stocks. The most important concentration of peatlands in the Nile Basin are the Sudd wetlands with 15,780 km^2 (about 50% of the total peatland area and 37% of the total carbon stocks of the entire Nile Basin) which remain largely undisturbed although they would be threatened by any reinitiation of the Jonglei canal drainage scheme.

Future Conflicts

In recent years, as we have seen, many wetlands throughout the Nile Basin have come under extreme pressure as government policies, socio-economic change and population pressure have stimulated the need for more agriculturally productive land. However, the drainage and cultivation of wetlands have had serious impacts on the regulatory functioning of the wetlands. This has led to a reduction in the storage and flow of water through the wetlands, with the result of serious negative impacts on communities downstream. At national levels many hydrological interventions exist, or are planned, throughout the Nile Basin to increase economic benefits and in particular, food security. However, these interventions can have numerous, possibly unpredictable consequences so that it is important to evaluate both costs and benefits. For example, a very likely consequence of flow regulation by building dams is the reduction of downstream seasonal flooding. This may inhibit the sustainable development of wetland ecosystems in the downstream region that rely on seasonal flooding for irrigation, although in order to assess this requires information describing their functioning and identifying the threats to these ecosystems at multiple spatial and temporal scales.

International disagreements over the management of wetlands have the potential to become one of the key conflict areas in many parts of the Nile catchment. The conflicts arise either because the use of the wetland in a particular area is negatively impacted by changes in water use upstream or by smaller groups of stakeholders implementing uses which negatively impact the wetland community as a whole. Solving these issues depends on the development of conflict resolution processes which require an understanding of issues across the whole river basin. There is, therefore, a need to develop local wetland management practices that provide resources for local crop cultivation alongside wetland conservation, while retaining the range of benefits to the regional community as a whole. Large scale changes in land-use and water-use as well as changes in climate, all impact wetland functions and services but management decisions usually focus on individual wetland projects and local site conditions. Dixon and Wood (2003) have highlighted the need to match the local community and hydrological needs of a region that can be achieved by sustainable wetland utilisation. They suggest that there is a clear gap between the traditional research expertise from local wetland projects and its translation to larger scales. The gap hinders the development of practical management and policy solutions that are needed to limit further wetland deterioration and associated losses of valuable ecosystem services. The mismatch between scales at which hydrological changes take place and the scale at which observations and predictions are made has to be resolved in order to increase our understanding of how hydroclimatic and landscape-driven changes impact landscape functioning. Large scale analyses require a combination of ground-based measurements, analytical modelling and statistical approaches as well as the use of remote sensing techniques, which are synthesised and interpreted using GIS-based approaches.

Sharing Water

There is a long history of conflict over the availability of the Nile's water across the Nile Basin. Most recently the main area of focus has been on the Grand Ethiopian Renaissance Dam (GERD) straddling the Blue Nile and constructed in Ethiopia. The GERD is Africa's largest hydropower plant and it dams the Blue Nile coming from Ethiopia's highlands just before it crosses into Sudan, where after merging with the White Nile it continues northwards to Egypt. For Ethiopia the dam offers a solution to its severe power problems. However, Egypt relies on the Blue Nile for freshwater supplies and as a consequence it opposes the project. Sudan is also affected, although less severely, and it will benefit from the supply of additional Hydro-electricity from the dam. The construction of the Grand Ethiopia Renaissance Dam, which was first proposed as long ago as 1924, began in 2011 after a period of relative political stability and economic growth which meant that Ethiopia was able to finance it with international support.

Conflicts over sharing of the Nile's water goes back to early in the 20th century. After Britain occupied Egypt in 1882 Britain's textile industry became increasingly dependent on Egyptian cotton. The cotton production was dependant on irrigation water from the Nile. Gamel Abdel Nasser, Egypt's prime minister between 1954 and 1956 and the president until 1970 decided to build a dam at Aswan, creating Lake Nasser, 200 km in length and extending from Egypt into Sudan. It was developed to end Egypt's dependency on upstream countries by storing the Nile waters in Egypt to end the periodic flooding and also bring hydro-electricity. The vision was of a dam that would store enough water to see Egypt through the worst drought in a century and smooth the river flow. Its construction was opposed by Sudan because of the social and environmental costs. Sudan favoured the Century Storage Scheme including a series of small dams in Uganda and Ethiopia together with the Jonglei Canal. The intention of the Jonglei Canal, at 360 km in length was to redirect the flow of the White Nile to bypass the Sudd swamps in South Sudan, where vast amounts of water is lost by evapotranspiration. The construction of the canal started in 1977 but was suspended in 1983, 100 km short of completion, due to the civil war. South Sudan's government is now considering cooperation with Egypt to complete it although the international community are less convinced that it is environmentally acceptable. The projected negative environmental impact of the canal has been a cause for controversy ever since it was first proposed. Extensive environmental impact assessments were commissioned, the most detailed being reported in *The Jonglei Canal: Impact and Opportunity*, edited by Paul Howell, Michael Lock and Stephen Cobb and published in 1988. They showed that there would be massive reductions in the extent of the wetlands and that the reduced evapotranspiration from the area could have major impacts on local climate and the ecosystem services provided to the local pastoralist communities.

Developing policies to govern the allocation of water resources to the countries of the Nile Basin requires accurate data on how much water there is in terms of sources and sinks. Unfortunately, ground observations are limited and often inaccurate due to the costs of maintaining these measurements and the geographical isolation of these sites. In fact, numbers of functioning hydrological monitoring stations have declined significantly over the last half century. However, the situation is now changing largely due to recent advances in hydrological satellite observations which now enables the collection of more reliable information. Hydrological satellites circling hundreds of miles from the earth's surface make recurring visits to the same sites several times a month. One of these is NASA's joint satellite mission Gravity Recovery and Climate Experiment. Satellite observations have been used to determine the total available water storage in the Nile Basin between 2002 and 2020 (Hasan et al., 2021). Overall, the data revealed that the total available water storage in the Nile Basin, from all sources, could reach an average of 180 billion cubic metres per year. The data also allowed an estimate of the two main sources, Lake Victoria and the Blue Nile basin and the two sink regions where water is lost by evapotranspiration, the Sudd wetlands and the main Nile area across Egypt. Data showed that the Lake Victoria source is about twice that of the Blue Nile Basin and the Sudd loses about twice the water loss by the main Nile region. The satellite data also confirmed that between 2002 and 2020 the Nile basin experienced much wetter conditions. This is informative in planning for more water volumes during the flood seasons. Despite the additional water it is still estimated that the region is 'water stressed' based on the availability of water per person per year; less than 1,000 cubic metres a year. If available water becomes less than 500 cubic metres a year to meet all water demands a region is considered to be under absolute water scarcity conditions. Because of current booming population projections to reach as many as 800 million by 2050 the basin is considered to be under severe water stress conditions. The need is therefore to get the 11 countries in the basin to agree water sharing plans to avoid chronic water shortages in the future. The key to cooperation will be good data sharing and close technical cooperation between the riparian states.

Ethiopia needs GERD's electricity to lift millions of its citizens out of poverty. It is expected to double Ethiopia's electricity generation capacity and provide energy to millions of people. But Egypt is concerned by GERD's impacts on water available for its agriculture which depends completely on Nile water and Sudan also sees potential risks for its own water supplies. The issue is that Egypt has been in virtual control of the Nile water for centuries and now it is understandably alarmed by Ethiopia's challenge to its domination of the basin. Mediation talks to agree GERD's operation have been ongoing for years but are currently stalled. As of April 2025 the construction of the dam was 98.66% complete. Filling occurs during the rainy season of July and August and the fifth and last filling was completed in October 2024. The project will create a reservoir with a volume of more than 74 billion cubic metres. However, while large-scale diversion upstream will have impacts on food production, as the GERD will generate hydropower it only stores water and this does not directly affect supply to Egypt. Nevertheless, agreement is needed on how to operate the dam if there is a prolonged period of dry years. However, it is argued that GERD could eventually help Egypt save water as the regulated flow will allow the Aswan dam to reduce its level, reducing evaporation by about 10%. In addition, Sudan will benefit from expanded irrigation and a share of the cheap electricity generated. However, Sudan has said it fears that Ethiopia's dam will impact on its own hydroelectric installations. Sudan has constructed a series of dams along its length of the Nile and they argue that river flow fluctuations caused by upstream dams can have major implications for their own downstream dams. This is also restricting the flow of water to Egypt, giving greater urgency for the need to resolve the long-running disputes on the distribution of water for the Blue Nile supplies to Sudan and Egypt. Absence of binding agreements on water flow control could certainly threaten the effectiveness of these dams and completion and operation of the GERD will require careful management of regional tensions and water security concerns.

Future Impacts of Climate Change

Climate trends in the Upper Nile Basin since the 1960s show that hot and dry years have become more common and models show that this is likely to continue, resulting in crop failures for subsistence farming, food shortages and humanitarian crises. Models of future climate of the Nile basin indicate that by the end of the century the amount of rain falling in the upper Nile Basin could increase by up to 20%. However, the projections also show that as well as increasing rainfall there will be more extreme events of hot and dry spells which will have severe impacts on human populations, their crop production and the ecology of the Nile wetlands. It is estimated that by late 21st Century, the frequency of hot and dry years could increase by a factor of 1.5 to 3.0. In the past, a hot dry year occurred about every 20 years but with the increase in frequency in the future a hot dry year could occur every 6 to 10 years. Large deficits in water flowing into streams and rivers from reduced rainfall during hot and dry years will amplify these effects. Hot and dry conditions can kill crops, reduce hydropower and reduce overall water availability. At present around 10% of the basin's population faces chronic water scarcity due to the region's seasonal aridity and the highly unequal distribution of water resource but by 2040 the number of people facing water scarcity could reach 35% of the population and a hot and dry year could push 45% of the people in the Nile basin (nearly 110 million people) into water scarcity. Under these circumstances a situation may arise where the downstream states of Egypt and Sudan will try to assert control over the Nile flow, while upstream states try to find ways to secure water for their own wellbeing. The need to establish equitable water allocation under ever worsening conditions of availability has led to predictions of water wars.

Papyrus Wetlands for Nature-based Solutions and Bioenergy Production

In recent years nature-based solutions (NbS) have been championed as actions to protect, sustainably manage and restore natural ecosystems to provide solutions for dealing with a range of global environmental challenges such as climate change, loss of biodiversity, pollution of water systems, the need to increase soil carbon sequestration, and flood regulation (Seddon et al., 2020). NbS are determined by the natural functions of wetland ecosystems that, for example, sequester carbon in their soils or remove contaminants from water. The potential of wetlands to supply multiple ecosystem services means that they can make important contributions to the provision of NbS (Xu et al., 2020; Ferreira et al., 2023). Thorslund et al., (2017) suggest that the relevant scale to assess wetland functions for NbS is at the landscape scale. This is because critical ecosystem services emerge from the aggregated effects of individual wetlands interacting with their surrounding landscape, which is referred to as the 'wetlandscape'. Climate change and large-scale land-use changes impact at the wetlandscape scale. Management and policy decisions therefore need to be made at the wetlandscape scale which is limited by the borders of the entire interconnected hydrological catchment. At the wetlandscape scale Nature-based Solutions have in common the requirement to be designed, implemented, managed and monitored by a partnership with local communities through a process that fully respects and champions local rights and knowledge, and concurrently generates local benefits. End users must be convinced of the benefits derived for them in managing the wetlands for Nature-based solutions.

The recognition of the valued, and in some cases essential, ecosystem services provided by papyrus wetlands has led to proposals that not only can actively growing papyrus wetlands be exploited for Nature-based solutions but it might be possible to reverse the continued decline in the areas of these wetlands to provide the extensive areas required to benefit from NbS. In order to contribute to the large scale requirement for NbS there would need to be mass replanting of papyrus or the clearing of suitable wet areas for the germination of papyrus seedlings (Boar, 2006, Terer et al., 2012). There have been some attempts to re-establish papyrus wetlands that have been damaged or completely removed and they suggest that it might be feasible to re-establish papyrus given favourable conditions. For example, Kiwango ėt al. (2013) reported the establishment of two small papyrus wetlands in villages on the shores of Lake Victoria near Rubondo National Park in the south west region of the Lake. The aim was to help the local community to replenish the fish stocks in the lake and to provide additional services provided by the papyrus. Two years after the wetlands were created the above ground biomass was close to that of pristine papyrus in the area, and Tilapia fish numbers had increased, which were the main benefits to the community. It was additionally anticipated that papyrus harvesting for fuel or fodder would be possible in the near future, if sustainable harvesting practices were adopted. In this study the importance of maintaining long term stability of water levels was emphasised although ironically the short term fall in lake levels between 2000 and 2005 was advantageous in exposing a detritus substrate and seedbank for establishing new papyrus stands from seed germination and seedling growth. This illustrates that short term falls in water levels may in fact offer the opportunity for papyrus wetlands to regenerate as long as water level recovery follows quickly enough to allow the rhizome mat to develop and float on the rising water.

There have also been suggestions that the future management of existing and re-established papyrus wetlands could involve the sustainable harvesting of papyrus as a source of fuel. If this is managed to allow for the continued regrowth of the harvested material then the NbS discussed above would still be retained, achieving a win-win situation. This would involve the regular removal of large areas of papyrus aerial growth and its subsequent recovery as new culms develop. However, the possible negative consequences of this for the ecological functioning of the papyrus swamp have raised concern, particularly because there have been few studies where the regular harvesting of the swamp has been monitored. Terer et al., (2012) investigated the effects of harvesting on aerial biomass and culm regeneration of papyrus growing in an undisturbed wetland at Lake Naivasha, Kenya. They harvested the papyrus culms at 6 and 12 monthly intervals over a period of 3 years and concluded that the that swamps should be zoned into 12 rotationally harvested plots, each harvested once a year, while also leaving a good proportion of the swamp undisturbed, to achieve a sustainable harvesting regime. They suggested that this harvesting regime would maintain biomass regeneration and health in terms of height and diameter of culms which would sustainably support local harvesting activities and avoid reduced vigour that occurs after more frequent harvesting. The undisturbed zones would remain as a permanent habitat for fauna, especially a group of papyrus specialist birds such as the papyrus gonolek. In addition, it would ensure that the natural functioning of the swamp in intercepting sediments and in buffering increased nutrient loads is maintained.

Current Bioenergy from Papyrus

Sub-Saharan Africa governments are challenged with finding mechanisms to stimulate development that provides jobs and education, alleviates poverty and hunger and provides sustainable energy while also protecting their natural resources and environments. Ensuring clean and affordable energy is an essential enabler of economic development and will lift people out of poverty and also reflects one of the key United Nations Sustainable Development Goals (no. 7) – ensuring access to affordable, reliable and sustainable modern energy for all, and adopted by all UN Member States in 2015. The primary source of energy for over 80% of the households of Sub-Saharan tropical Africa is biomass, predominantly in the form of charcoal or wood, which is used mainly for cooking and heating. Fuel-wood remains the main source of domestic energy for the rural poor but charcoal is the major source of energy for the urban poor and even though the share of the energy source produced by biomass is declining, the numbers dependent on it continues to grow. An estimated 40% rise in the demand for bioenergy by 2040 will increase the pressure on the forest biomass stocks, with efforts to promote more sustainable wood production hindered by the operation of much of the fuel-wood and charcoal supply chain outside the formal economy (Gronau et al., 2018). The supply of fuel-wood to produce charcoal or for burning directly is obtained from a wide range of sources ranging from scrubland and savannah to forest ecosystems. In many areas, the supply of this wood is not sustainable and its continued use has led to the depletion of forest ecosystems in particular, and it has been shown that in many areas with the current levels of biomass harvesting reserves of fuel-wood could be totally exhausted within the next decade or two (Wessels et al., 2013). It is therefore essential to identify new forms of biomass feedstock that can be produced and supplied sustainably and would reduce the current dependence on unsustainable forms of energy.

A potential source of bioenergy is non-woody natural vegetation which could be harvested on an annual basis and then regrows. However the widespread view is that the sustainability of this in Sub-Saharan Africa, where all productive land is required for growing food crops, is questionable. There have also been proposals to plant 'marginal land' that may be sustainably used to produce dedicated energy crops and another approach has been to use feedstocks from agricultural waste (Eluwa and Kilanko, 2024). However both of these approaches have met with opposition because of concerns about conflicts with food production. This is because the vast majority of the population are primarily dependant on agriculture as a means of survival and many of them face growing seasons when they are unable to produce enough crops to satisfy the demands of the indigenous population. In some countries concerns around food security have resulted in governments actively cautioning against the development of bioenergy and preventing a situation when fuel production is seen as aggressively displacing food production. It has, in fact, become increasingly recognised that the growing of energy crops in Sub-Saharan Africa may have a number of additional negative environmental and social impacts including increased greenhouse gas emissions, deforestation, biodiversity loss, soil erosion, soil carbon loss, unsustainable water use, conflicts over land-use rights and spikes in staple food-crop prices.

In the search for alternative sources of biomass for biofuel production, wetlands and in particular papyrus-dominated wetlands, have been largely overlooked. Given its potential as a high yielding feedstock, there is growing interest in the use of papyrus biomass in the production of biofuels (Jones et al., 2016). However, at the present the use of papyrus as a fuel is very limited. Papyrus culms can be burnt in open fires and in stoves for cooking and heating but it needs to be dried and compressed for effective heat production. There are several forms in which it can be efficiently combusted, including chips, pellets and briquettes. Making pellets and briquettes upgrades the fuel and facilitates utilisation by increasing bulk density, stabilising moisture content and reducing dust emissions in the handling of the fuels with the result that it leads to a more stable, uniform product with more efficient combustion control. Efforts were made in the 1980s to establish the production of papyrus briquettes as an alternative to wood charcoal. A pilot factory was established near Kigali, Rwanda but the project did not succeed due the failure to commercialise the production of briquettes (Jones, 1984). More recently, the Future from the Field (FfrF) project led by the Massachusetts Institute of Technology (MIT) proposed that the small-scale production of carbonised papyrus briquettes could be used to incentivise the sustainable management of papyrus vegetation (Morrison et al., 2014). The FfrF process is considered to require minimal technical expertise and can utilise locally available materials which can be purchased locally. It has been suggested that a worker can produce enough briquettes to pay for the equipment and start making a profit in less than a month.

In the FfrF process carbonised papyrus is produced using a converted 200 Litre oil drum as a combusting kiln. A large hole in the top of the drum and a number of smaller holes in the base allow for airflow during the initial combustion of dried papyrus culms. The majority of the volatile organic compounds in the papyrus burn off during the initial stages of combustion. Once the papyrus reaches carbonising temperature (~450°C) the kiln is sealed to create an anaerobic environment to produce charcoal. The cooled carbonised papyrus is crushed and mixed with 5–10% by weight casava flour in heated water to produce a porridge-like consistency, which can be formed into cuboid briquettes. The metal press for the production of briquettes can be easily manufactured locally and can be operated by a single person (Morrison et al., 2014).

In another study on the impact of papyrus briquette production on rural livelihoods and forest resources Gronau et al. (2018) used a computable general equilibrium (CGE) model and data from a rural village in Zambia to ask 'does papyrus harvesting and processing improve local livelihood?' and 'does papyrus harvesting and processing have the potential to reduce pressure on local forest resources?' This was the first study to assess the impact of papyrus briquettes on rural livelihoods and use of forest resources at the village level in Sub-Saharan Africa. The model showed that papyrus briquetting and use as a biofuel not only improves the local livelihoods but the households tend to give up their forest harvesting and charcoal production activity in order to switch to papyrus use. As a consequence, forest land is conserved and is replaced by a smaller area for bioenergy generation. Papyrus briquettes thus substitute energy supplies from firewood and charcoal and have the potential to reduce pressures on local forest resources in the region. Furthermore, users state that they prefer briquettes to firewood and charcoal. The main advantages being longer and stronger heat, clean burning and less smoke, resulting and health improvements (less coughing and lung damage as a result of smoke inhalation). However, in order to meet demand for supply of briquettes there is a need for more training in the process of manufacture and improved marketing support to increased production (Gronau et al., 2018).

Future Bioenergy from Papyrus

As we have seen, all efforts to date for generating energy from papyrus have utilised solid fuel (compressed plant material) but plant biomass can also provide energy in liquid and gaseous forms (Jones et al., 2016). These are more flexible in their utilisation, particularly as transport energy, as well as being more energy intensive and cleaner burning than solid biomass (Dasappa, 2011, Eluwa and Kilanko, 2024). The three most important bioenergy conversion routes are (i) direct combustion of solid fuels, (ii) enzymatic saccharification followed by fermentation to produce liquid bioethanol (used primarily as a transport fuel in East Africa; Miftah and Mutta, 2024) and (iii) anaerobic digestion to produce gaseous biomethane, which can be used for transport, although the primary market for it in Sub-Saharan Africa is likely to be fuel for stoves as a replacement for wood and charcoal. The liquid fuel and biogas are referred to collectively as 'biofuel'. The biomass feedstocks for the conversion routes to biofuels are referred to as first and second generation, where first generation feedstocks are currently cultivated as food crops and second generation feedstocks are non-food crops. At present, most of the first generation feedstocks for biofuel production are produced from two crops; sugarcane (*Saccharum officinarum*) and maize (corn) (*Zea mays*). Liquid bioethanol is produced by, respectively, the fermentation of sugar extracted from sugarcane or derived from the hydrolysis of starch in maize. Understandably, there is considerable concern about using first generation feedstocks for biofuel production and diverting it from the food supply chain and it is very unlikely that these feedstocks will be acceptable in Sub-Saharan Africa where food shortages are a regular occurrence. It is essential therefore to identify second generation feedstocks that might support a bioenergy industry on a regional scale and papyrus clearly meets this requirement as its production does not compete for land with food crops when utilised sustainably.

The production of bioethanol and other fermentation products from papyrus biomass is a multistep process that normally includes pre-treatments, enzymatic hydrolysis and yeast fermentation. Pre-treatment aids the recovery of cellulosic content, a key component of plant cell walls from lignocellulosic biomass and renders it digestible to cellulases, a group of enzymes that break down cellulose into simpler sugars like glucose. These reducing sugars are then converted to ethanol by fermentation. Currently, the total cost of bioethanol production from lignocellulosic feedstock is still much higher than production from starch or sugar-based feedstock. This is because production of bioethanol from lignocellulosic feedstock requires a larger investment in more diverse enzymes to first convert plant cell walls to sugar than is needed to simply release sugar from starch in first generation feedstocks. Although enzyme costs have decreased in the last few years, the high cost of enzyme production and the requirement for higher enzyme dosage for hydrolysis of biomass are the main hurdles for the current economic viability of lignocellulosic bioethanol. In addition to cost, enzyme diversity is an issue because the plant cell wall, with its many polysaccharides, is far more complex than starch. To make full use of plant cell walls, cocktails of enzymes capable of synchronised digestion of the polymers are needed.

Another source of biofuel is the process of anaerobic digestion; the decomposition of organic matter in an oxygen-free environment to produce biogas that is usually made up of around 60% methane and 40% CO_2. The process of anaerobic digestion involves the production of biogas from fatty acids, amino acids and sugars. Through the process of acidogenesis, complex organic matter is broken down into simpler molecules by the action of bacteria in the digestate. Ammonia, hydrogen, carbon dioxide and hydrogen sulphide are the main products of this process. The products are further processed during acidogenesis when acetic acid, carbon dioxide and water are formed. Methanogens then act on these products to form methane. The technology of biogas production is well established and can be implemented in simple set-ups for single households or small communities and as part of bigger complex plants feeding a larger grid system.

Future long term developments to derive greater value from the processing of papyrus could see the development of biorefineries. A biorefinery is a facility that integrates biomass conversion processes to produce bioethanol as well as value-added chemicals from biomass. It is analogous to a petroleum refinery which produces multiple fuels and products from petroleum. In this process a raw material consisting primarily of polysaccharides and lignin enters the biorefinery and it is fractionated and converted into a mixture of products, mainly bioethanol, but also about 5% goes to high value chemical products (Ragauskas et al., 2006). In the development of what is known as the 'lignin first' methods the aim is to preserve lignin for subsequent conversion to a range of useful, high value, products. In order to do this a detailed analysis of the chemical composition of the plant components needs to be carried out and recently Rosado et al. (2021 and 2022) have for the first time analysed the differences in the content, composition and structure of lignins from the culm 'rind' and pith of papyrus. The analyses revealed that the lignins from these components of the of papyrus culms differed greatly not only in their contents but also in composition and structure, important information for developing the fractionation process. Fatty acids, sterols and other aromatic compounds can also be extracted from *Papyrus*. The exploitation of these low-volume-high-value chemicals can make an important contribution to the economic return on developing biomass fuels.

A further opportunity for the large-scale utilisation of harvested papyrus is in the manufacture of bio-composite materials. Bio-composites use papyrus fibres combined with natural or synthetic binders, in the form of polymers or resins, to produce a range of products including insulation panels and other construction materials, and biodegradable packaging material. Research on papyrus composites is needed to improve the fibre properties by physical and chemical treatments and to improve the strength and durability of papyrus-based bio-composite construction materials.

Maintaining Wetland Carbon Stocks

National commitments to reduce carbon emissions made by parties to the United Nations Framework Convention on Climate Change (UNFCCC) and the associated Paris Agreement have focussed attention on preserving the large stocks of carbon in peatlands of the tropics.

According to the Nile Basin Initiative corporate report 2020 (NBI, 2019), the total area of peatlands and other organic soils in the Nile Basin amounts to about 3 million hectares. Because of uncertainties about the depth of this peat estimates of the amount of carbon stored in the peatlands are very difficult to make but the area is estimated to contain a carbon stock of between 4.4 and 10 Gt C. The most important concentration of peatlands in the Nile Basin is in the Sudd wetlands, with an estimated area of organic soils of 1.6 million hectares. This accounts for 50% of the peatland area and 37% of the total C stocks in the entire Nile Basin. Other substantial peat deposits in the Nile Basin are located in the Albert Nile swamps, the Lake Kyoga swamps in Uganda and the Kagera wetlands in Rwanda. There is increasing concern about the environmental consequences of large-scale agricultural conversion of these peatlands when they are drained to allow establishment of crops or the harvesting of peat for energy production. For example, Rwanda, one of the Nile Basin countries, is currently planning to exploit its peatland resources in Gisagara District and to construct an 80 MW peat to power plant. The development alongside the Akanyaru River involves the construction and operation of both peat harvesting and power generation facilities. The project is expected to increase national power capacity by 40% and increase the countries' energy independence.

The widely recognised role of peatland carbon storage in climate change mitigation has led to proposals that tropical peatlands are managed to protect the underlying carbon stocks. Currently, when wetlands are drained for cultivation of crops and exposed to the atmosphere the carbon content is rapidly lost through the process of oxidation, dissolution and in some cases fire. High rates of decomposition and compaction also cause the peat surface to subside below its original elevation and there is a rapid loss in soil fertility. However, there have been suggestions that it is possible to design and implement sustainable approaches that allows crop production while preserving peatland's ecological integrity, goods and services through what is termed 'paludiculture', loosely defined as the 'sustainable production of biomass on wet or rewetted peatlands' (Tan et al., 2021). It merges traditional agricultural know-how with new practices when using wet peatland and preserving the peat body and its characteristics. It has been suggested that the design of comprehensive paludiculture strategies offers a promising opportunity for the sustainable use of peatlands in the Nile Basin and the Equatorial Lakes Region. Sustainably using peatlands is aimed at maintaining a neutral carbon balance, with the long-term prospects of turning these ecosystems back into carbon sinks. There are a relatively small number of crops that can be cultivated under wetland agriculture and this suggests that paludiculture may have limited scope. However, suitable crops that are currently grown in wetlands of the Nile Basin are cocoyam or taro (*Colocasia esculenta*), sago palm (*Metroxylon sagu*) and rice (*Oryza sativa*), which have the potential to provide economic benefits while helping to maintain the ecological functions of peatlands.

Wastewater Treatment by Papyrus Wetlands

Wetland ecosystems are being used increasingly worldwide for improving water quality and disposal of wastewater because in the right location they have been recognised as low cost and effective treatment systems (Perbangkhem and Polprasert, 2010). The highest value of ecosystem service in papyrus dominated wetlands has been attributed to wastewater treatments in locations where urban developments are releasing this wastewater into Lake Victoria wetlands (Emerton et al., 1998). Good performance of wetlands for wastewater treatment depends on the growth potential and ability of macrophytes to develop sufficient root systems for microbial attachment and material transformations, and to incorporate nutrients into the plant biomass that can subsequently be harvested to remove the nutrients. A high below-ground biomass productivity favours more uptake and translocation of nutrients to the growing aerial parts.

The Nakivubo swamp in the Inner Murchison Bay that connects the city of Kampala (Uganda's capital) to Lake Victoria has been receiving partially treated sewage from Kampala for more than 40 years. Two distinct macrophyte zones in the swamp are dominated by either papyrus or *Miscanthidium violaceum*. Papyrus has been demonstrated to be more effective than other aquatic macrophytes, including *Miscanthidium* and *Phragmites,* at removing excess nutrients from wastewater due to its thin and loose root mat that allows efficient water-plant interaction, whereas the thick and compact root mat of *Miscanthidium* and the sediment that the roots of *Phragmites* are embedded in, severely restricts access to the flowing wastewater and reduces their efficiency of nutrient removal (Azza et al., 2000; Kansiime et al., 2007). In addition, higher rhizome and root productivity of papyrus resulted in greater uptake and translocation of nutrients to the growing parts of the plant and higher aerial productivity. To avoid nutrient recycling as the papyrus dies at the end of its growing cycle, and to achieve effective removal of the nutrients, the periodic harvesting of the aerial vegetation is necessary. Wastewater purification therefore not only provides a very valuable regulating ecosystem service but it also adds to the value of the provisioning service of harvestable biomass production.

In the Lake Victoria basin, lake fringe swamps and other wetlands are increasingly used to treat secondary domestic waste water effluents from adjacent towns as well as surface runoff and raw sewage from local communities. However, as these swamps are encroached for agricultural crop production their effectiveness is reduced to the point where alternative approaches need to be adopted which revolve around the development of constructed wetlands (Kyambadde et al., 2005). This entails constructing an artificial wetland to grow the macrophyte within a contained environment on a bed of gravel at the bottom and sandy loam on top, and directing the flow of contaminated water through the system. Different designs of constructed wetlands have been developed where the flow of water is horizontal, either through a rhizome mat and rooted substrate or below a floating mat with free contact of the root with the water. Kyambadde et al. (2004) showed that horizontal sub-surface flow systems have limited ability to remove phosphorus because the root mass fills the pore spaces of the substrate, diverting flow to lower in the substrate. Roots are limited in their contact with the wastewater, therefore limiting the rootzone oxidation, absorption and plant uptake. The role of vegetation used in these systems have been difficult to elucidate, although in most trials *C. papyrus* appears to be more effective than other species, possibly because it has such high growth rates. Floating treatment wetlands (FTW) are the newest artificial wetland systems developed to treat water and wastewater. In FTW, plants grow as a floating platform, the substrate is discarded, and their mats are free in the water column taking nutrients from water. They therefore have a higher pollutant removal efficiency than constructed wetlands in which vegetation is rooted in sediment (Sanchez-Galvan et al., 2022).

Conclusion

Although it is now abundantly clear that papyrus wetlands have the potential to support rural livelihoods, while at the same time contributing to water quality control, climate regulation and conserving biodiversity, their distribution in eastern and southern Africa is continuing to decline. Degradation and associated loss of ecosystem services affects the poorest, land dependent communities that are most directly dependent on wetland resources and functions. This chapter has reviewed prospects for reversing this decline in the future to ensure that papyrus wetlands continue to make a significant contribution to the sustainable ecology of the Nile Basin region.

While several research projects have identified, at the local scale, threats and potential solutions to protect and even restore papyrus wetlands in the future, there has been a lack of coordination to integrate this information and solutions at the regional scale. An attempt to overcome this problem has been the establishment of the Nile Basin Initiative with a mission to achieve sustainable socio-economic development through the equitable utilisation of the common Nile Basin water resources in the 11 member states. In addition, the NBI is complemented by the Ramsar Convention on Wetlands, an international treaty that aims to conserve and manage wetlands worldwide by national action and international cooperation. The convention offers guidance on sustainable use of wetlands, balancing human needs with conservation goals.

Scenarios for the future of papyrus wetlands range from community-based natural resource management of common ownership areas of wetlands that balances livelihood needs with resource conservation for ecosystem services (Gosling et al., 2017) to large scale programmes that utilise the papyrus wetlands for nature-based solutions to address the global issues of water pollution, climate change, biodiversity loss and energy poverty. In terms of societal needs the community-based management of papyrus wetlands has the opportunity to provide the most valuable ecosystem services but communicating these benefits is difficult, with the result that they are most frequently ignored. However, there is evidence that a strong and well organised community approach can be locally supported when benefits are clear, widely available and the costs of management are fully compensated. With respect to large scale programmes, if some of the extensive areas of papyrus wetlands such as the Sudd Swamps in South Sudan and Kyoga Lake Complex in Uganda are to be conserved then their nature-based services, based on their natural functions, need to be identified and valued. These services include flood regulation, carbon sequestration and biomass production. These areas of wetland, because of their size, are in the position to make globally significant contributions to climate control and bioenergy production in the future.

References

Azza, N.G.T., Kansiime, F., Nalubega, M. and Denny, D. (2000). Differential permeability of papyrus and *Miscanthidium* root mats in Nakivubo swamp, Uganda. *Aquatic Botany*, 67, 169–178.

Boar, R.R. (2006). Responses of a fringing *Cyperus papyrus* L. swamp to changes in water level. *Aquatic Botany*, 84, 85–92.

Carter, G.S. (1955). *The papyrus swamps of Uganda*. Cambridge: W. Heffer & Sons.

Dasappa, S. (2011). Potential of biomass energy for electricity generation in Sub-Saharan Africa. *Energy for Sustainable Development*, 15, 203–213.

Dixon, A.B. and Wood, A.P. (2003). Wetland cultivation and hydrological management in eastern Africa: Matching community and hydrological needs through sustainable wetland use. *Natural Resources Forum*, 27, 117–129.

Eluwa, S. and Kilanko, O. (2024). Biofuel as an alternative for Sub-Saharan Africa's transition to cleaner energy. *Academia Green Energy*, 1(1). doi:10.20935/AcadEnergy6227.

Emerton, L., Iyango, L., Luwum, P. and Malinga, A. (1998). *The present economic value of Nakivubo urban wetland, Uganda*. Nairobi: IUCN Eastern African Regional Office.

Farmer, J., Langan, C. and Smith, J.U. (2022). Temporal variability in heterotrophic carbon dioxide emissions from a drained tropical peatland in Uganda. *Frontiers in Soil Science*, 2, 904647.

Ferreira, C.S., Kašanin-Grubin, M., Solomun, M.K., Sushkova, S., Minkina, T., Zhao, W. and Kalantari, Z. (2023). Wetlands as nature-based solutions for water management in different environments. *Current Opinion in Environmental Science & Health*, 33, 100476.

Gosling, A., Shackleton, C.M. and Gambiza, J. (2017). Community-based natural resource use and management of Bigodi Wetland Sanctuary, Uganda, for livelihood benefits. *Wetlands Ecology and Management*, 25, 717–730.

Gronau, S., Winter, E. and Grote, U. (2018) Papyrus, forest resources and rural livelihoods: A village computable general equilibrium analysis from northern Zambia. *Natural Resources*, 9, 268–296.

Hasan, E., Tarhule, A. and Kirstetter, P.E. (2021). Twentieth and twenty-first century water storage changes in the Nile River basin from GRACE/GRACE-FO and Modelling. *Remote Sensing*, 13, 953. doi:10.3390/rs13050953.

Howell, P., Lock, M. and Cobbs, S. (eds). (1988). *The Jonglei Canal: Impact and opportunity*. Cambridge: Cambridge University Press.

Jones, K., Lanthier, Y., van der Voet, P., van Valkengoed, E., Taylor, D. and Fernández-Prieto, D. (2009). Monitoring and assessment of wetlands using Earth Observation: The GlobWetland project. *Journal of Environmental Management*, 90, 2154–2169.

Jones, M.B. (1984). Biofuel developments in Central Africa: Technology transfer from Ireland. In J. Twidell, I. Hounam and C. Lewis (eds), *Energy for rural and island communities* (pp. 355–361). Oxford: Pergamon Press.

Jones, M.B., Kansiime, F. and Saunders, M.J. (2016). The potential use of papyrus (*Cyperus papyrus* L.) wetlands as a source of biomass energy for Sub-Saharan Africa. *GCB Bioenergy*, doi:10.1111/gcbb.12392.

Kansiime, F., Kateyo, E., Oryem-Origa, H. and Mucunguzi, P. (2007). Nutrient status and retention in pristine and disturbed wetlands in Uganda: Management implications. *Wetlands Ecology and Management*, 15, 453–467.

Kiwango, Y., Moshi, G., Kibasa, W. and Mnaya, B. (2013). Papyrus wetland creation, a solution to improve food security and save Lake Victoria. *Wetland Ecology and Management*, 21, 147–154.

Kyambadde, J., Kansiime, F. and Dalhammar, G., (2005). Nitrogen and phosphorus removal in substrate-free pilot constructed wetlands with horizontal surface flow in Uganda. *Water, Air, and Soil Pollution*, 165, 37–59.

Kyambadde, J., Kansiime, F., Gumaelius, L. and Dalhammar, G. (2004). A comparative study of *Cyperus papyrus* and *Miscanthidium violaceum*-based constructed wetlands for wastewater treatment in a tropical climate. *Water Research*, 38, 475–485.

Miftah, F.K. and Mutta, D. (2024). Potential markets and policies for sustainable liquid biofuel production with emphasis to Eastern Africa countries: A review. *Energy, Sustainability and Society*, 14, 1.

Morrison, E.H.J., Banzaert, A., Upton, C., Pacini, N., Pokorny, J. and Harper, D.M. (2014). Biomass briquettes: A novel incentive for managing papyrus wetlands sustainably? *Wetlands Ecology and Management*, 22, 129–141.

NBI. (2019). *Nile Basin Initiative corporate report 2019*. https://nelsap.nilebasin.org/index.php/en/media-items/documents-publications/28-nbi-corporate-report-2019-1/file.

Perbangkhem, T. and Polprasert, C. (2010). Biomass production of papyrus (*Cyperus papyrus*) in constructed wetland treating low-strength domestic wastewater. *Bioresource Technology*, 101, 833–835.

Ragauskas, A.J., Williams, C.K., Davison, B.H., Britovsek, G., Cairney, J., Eckert, C.A., FrederickJr, W.J., Hallett, J.P., Leak, D.J., Liotta, C.L. and Mielenz, J.R. (2006). The path forward for biofuels and biomaterials. *Science*, 311(5760), 484–489.

Rosado, M.J., Bausch, F., Rencoret, J., Marques, G., Gutiérrez, A., Rosenau, T., Potthast, A. and del Río, J.C. (2021). Differences in the content, composition and structure of the lignins from rind and pith of papyrus (*Cyperus papyrus L.*) culms. *Industrial Crops and Products*, 174, 114226.

Rosado, M.J., Marques, G., Rencoret, J., Gutierrez, A., Bausch, F., Rosenau, T., Potthast, A. & Del Río, J.C. (2022). Chemical composition of the lipophilic compounds from the rind and pith of papyrus (*Cyperus papyrus* L.) stems. *Frontiers in Plant Science*, 13, 1097866. doi:10.3389/fpls.2022.1097866.

Sanchez-Galvan, G., Olguin, E.J., Melo, F.J., Jimenez-Moreno, D. and Hernandez, V. J. (2022). Pontederia sagittate and Cyperus papyrus contribution to carbon storage in floating treatment wetlands established in subtropical urban ponds. *Science of the Total Environment*, 832, 154990.

Seddon, N., Chausson, A., Berry, P., Girardin, C. A., Smith, A. and Turner, B. (2020). Understanding the value and limits of nature-based solutions to climate change and other global challenges. *Philosophical Transactions of the Royal Society B*, 375, 20190120. doi:10.1098/rstb.2019.0120.

Tan, Z. D., Lupascu, M. and Wijedasa, L. S. (2021) Paludiculture as a sustainable land use alternative for tropical peatlands: A review. *Science of the Total Environment*, 753, 142111.

Terer, T., Triest, L. and Muthama Muasya, A. (2012). Effects of harvesting *Cyperus papyrus* in undisturbed wetland, Lake Naivasha, Kenya. *Hydrobiologia*, 680, 135–148.

Thorslund, J., *et al.* (2017). Wetlands as large-scale nature-based solutions: Status and challenges for research, engineering and management. *Ecological Engineering*, 108, 489–497.

Turyahabwe, N., Tumusiime, D.M., Kakuru, W. and Barasa, B. (2013). Wetland use/cover changes and local perception in Uganda. *Sustainable Agriculture Research*, 2, 95–105.

Wessels, K.J., Colgan, M.S., Erasmus, B.F.N., Asner, G.P., Twine, W.C., Mathieu, R., Van Aardt, J.A.N., Fisher, J.T. and Smit, I.P.J. (2013). Unsustainable fuelwood extraction from South African savannas. *Environmental Research Letters*, 8(1), 014007.

Xu, T., Weng, B., Yan, D., Wang, K., Li, X., Bi, W., Li, M., Cheng, X. and Liu, Y. (2019). Wetlands of international importance: Status, threats, and future protection. *International Journal of Environmental Research and Public Health*, 16, 1818. doi:10.3390/ijerph16101818.

Xu, X., Chen, M., Yang, G., Jiang, B. and Zhang, J. (2020). Wetland ecosystem services research: A critical review. *Global Ecology and Conservation*, 22, e01027.

Index

Index to *Papyrus and Wetlands Management in Africa*
References to illustrations are in *italics*.